HEALTHY ATOMS AND SICK ATOMS

PSYCHE GENESIS OF ELECTRONS AND NUCLEONS AND ATOMS

EZZAT E. MAJD POUR, M.D.

outskirts
press

Healthy Atoms and Sick Atoms
Psyche Genesis of Electrons and Nucleons and Atoms
All Rights Reserved.
Copyright © 2021 Ezzat E. Majd Pour, M.D.
v1.0

The opinions expressed in this manuscript are solely the opinions of the author and do not represent the opinions or thoughts of the publisher. The author has represented and warranted full ownership and/or legal right to publish all the materials in this book.

This book may not be reproduced, transmitted, or stored in whole or in part by any means, including graphic, electronic, or mechanical without the express written consent of the publisher except in the case of brief quotations embodied in critical articles and reviews.

Outskirts Press, Inc.
http://www.outskirtspress.com

ISBN: 978-1-9772-3770-5

Cover Photo © 2021 www.gettyimages.com. All rights reserved - used with permission.

Outskirts Press and the "OP" logo are trademarks belonging to Outskirts Press, Inc.

PRINTED IN THE UNITED STATES OF AMERICA

A NOTE FROM AUTHOR

Hundreds of New Findings that has been presented in this book, are Author's Discoveries and Inventions only. In past Human history no One has written about those research works, new creations and inventions.

The Author discovered that Electrons, Neutrons, Protons and Atoms have been constructed from Fundamental Particles, Particle Clouds, Particle Composites, and Particle Cloud Compound constructions of numerous and different kinds. The Fundamental Particle population of some of existing Nano Units exceed hundreds Billions of individual Fundamental particle population. Then calling the Electron, Neutron, Atom, Proton, etc. only one Fundamental Particle has no Merit.

Author discovered Fundamental Particle's Chemistry, The General chemistry of Fundamental Particles, The Organic chemistry of Fundamental Particles, and Biochemistry science of Fundamental Particles, as well as Author discovered Fundamental Particle's Biological sciences in multiple different fields, Particle Cloud Psyche genesis, Fundamental Particle Cloud Thought Current genesis, Atom's Particle Intelligence Systems, exogenous Particle Circulation systems, and indigenous Particle Cloud Circulation systems, Electrons and Nucleon's particle intelligence system centers, and Atom's Particle Circulation systems. Author discovered Molecular Evolution in

Fundamental Particles, Evolution in Elements, Evolution in Micro units (Cells and tissues), and Evolution in Macro unit structures and construction of Existing Things in Universe. Authors discoveries at present time are over hundreds, which no one has done these works before Author, at no time.

Also Author discovered the cause of the Particle transmitted diseases, causes of the Psychiatric diseases, and other numerous Particle Cloud transmitted disorders. Author discovered Plant's central intelligence system centers, and Plants peripheral intelligence system centers, Author discovered intelligence Genesis inside CNS Electrons and Nucleons, and knowledge accumulation phenomenon inside Atoms, through information image particle cloud storage system, under A. S. I. FP. Mol. CIC. which has utmost importance in teaching, learning, education, training, intelligence and knowledge genesis phenomenon. Author discovered Atom's psyche, Atom's PCS. and CNS Particle Cloud Circulation systems, Author discovered internal Electron intelligence system centers, and internal Nucleon intelligence system centers, and discovered hundreds of other nonpublished other discoveries, etc. which are going to be presented in numerous other books.

All of these works done by Author only. these Discoveries are not connected, extracted from others findings, or Creations.

REFERENCE: NONE

EZZAT E. MAJD POUR, M.D.

ADDRESS: ezzatmajd@icloud.com

April - 10 - 2021

The Author's discovery rights and intellectual properties in this book are protected and reserved only for him through the United States copyright office. the Owner of this Book's creations discoveries is Author only. No parts of this Book's discoveries must not be violated by any means by no one at no time, either through Printing, reproduction, distribution digitally or physically without written permission directly from the Author.

This book must not be reproduced, printed, stored, retrieved, copied, recorded, transmitted digitally or through physical printing modalities Without written permission from the author.

EZZAT E. MAJD POUR, M.D.

Address:ezzatmajd@icloud.com

THE SUBJECTS   PAGES

1 – THE FUNDAMENTAL PARTICLES OR UNITS OF THE MATTER

2 – THE FUNDAMENTAL PARTICLES COLONY

3 – EARTH'S FUNDAMENTAL PARTICLES, FUNDAMENTAL PARTICLE CLOUDS

4 – EARTH'S BIOFRIENDLY SENSIBLE PARTICLE CLOUDS

5 – LIGHT FUNDAMENTAL PARTICLES

6 – THE INFRARED LIGHT – THERMAL FUNDAMENTAL PARTICLES

7 – THE ULTRAVIOLET LIGHT – ELECTRIC FUNDAMENTAL PARTICLES

8 – UNIT OF QUANTUM LEVEL ELECTRIC ENERGY, BIOFRIENDLY NON DISCOVERED PARTICLES

9 – THE EARTH'S BIOHOSTILE FUNDAMENTAL PARTICLES

10 – THE EARTH'S BIOFRIENDLY SENSIBLE SONIC FUNDAMENTAL PARTICLES

11 – THE MEMORIZATION PHENOMENON

12 – NON SENSIBLE BIOFRIENDLY NON DISCOVERED FUNDAMENTAL PARTICLES OF EARTH

13 – QUANTUM ENERGY CONTENTS OF THE FUNDAMENTAL PARTICLES, QUANTUM ENERGY DONORS, QUANTUM ENERGY RECIPIENTS, PERFORMING QUANTUM WORKS INSIDE ELECTRONS, NEUTRONS, PROTONS

14 – THERMAL ATOMS, THERMAL ELECTRONS AND NUCLEONS, RED COLOR THERMAL PARTICLES

15 – THE THERMAL ENERGY FUNDAMENTAL PARTICLES, QUANTUM LEVEL THERMAL ENERGY UNIT.

16 – CHEMICAL INTERACTION DIFFERENCES BETWEEN DIFFERENT THERMAL PARTICLES

17 – THE ESSENTIAL LAWS OF FUNDAMENTAL PARTICLE CHEMISTRY, (A. S. I. FP. Mol. CIC.)

18 – INTRODUCTION TO FUNDAMENTAL PARTICLE BIOCHEMISTRY

19 – CHEMICAL COMBINATION BETWEEN PLANT'S TWO DIFFERENT FUNDAMENTAL PARTICLE MOLECULE

20 – THERMAL (ENERGY) ATOMS, AND ELECTRIC (ENERGY) ATOMS

21 – ELECTRIC FUNDAMENTAL PARTICLE ATOMS, ELECTRIC FUNDAMENTAL PARTICLE MOLECULES

22 – LIGHT PARTICLE ATOMS, AND LIGHT FUNDAMENTAL PARTICLE MOLECULES

23 – CONSTRUCTION OF NANO UNITS, CREATION OF ELECTRONS, NEUTRONS, PROTONS, ETC.

24 – HOMOGENOUS FUNDAMENTAL PARTICLE CONSTRUCTION ELECTRON, NEUTRON, PROTON, ATOMS

25 – SPECIFIC ENERGY CONTENT ELECTRONS, PROTON, NEUTRON AND ATOMS

26 – ENERGY (QUANTUM) STORAGE, AND (QUANTUM) ENERGY RETRIEVAL IN NANO –SITES

27 – BIOCHEMISTRY; COLOR OF PLANTS, PLANT GENESIS, COLOR GENESIS, TISSUE GENESIS, EVOLUTION.

28 – NONDISCOVERED BIOFRIENDLY FUNDAMENTAL PARTICLES, ENERGY PROVIDER PARTICLES, NANO SITES

29 – QUANTUM ENERGY PROVIDERS FOR DOING NANO TASKS AT QUANTUM LOCATIONS AND AT MICRO SITES

30 – LIGHT ENERGY PROVIDERS IN PLANTS, UNIT OF LIGHT ENERGY, UNIT OF ELECTRIC ENERGY FOR NANO LEVEL

31 – GENERAL FUNDAMENTAL PARTICLE CHEMISTRY, MAIN LAWS IN PARTICLE CHEMISTRY

32 – CHEMICAL COMBINATIONS BETWEEN EXOGENOUS FUNDAMENTAL PARTICLES (EXTERNAL ATOM PARTICLES) WITH INDIGENOUS FUNDAMENTAL PARTICLES (INTERNAL ATOM PARTICLE MOLECULES) COMPOUNDS

33 – THE A. S. I. FP. Mol. CIC. BETWEEN WEAK FUNDAMENTAL PARTICLE MOLECULES

34 - A. S. I. FP. Mol. CIC. BETWEEN INTERMEDIATE WEIGHT FUNDAMENTAL PARTICLE MOLECULES

35 – A. S. I. FP. Mol. CIC. BETWEEN HEAVY WEIGHT FUNDAMENTAL PARTICLE MOLECULES

36 – FUNDAMENTAL PARTICLE CLOUD CIRCULATION SYSTEMS (PCS) AND A. S. I. FP. Mol. CIC.

37 – NANO NEUROLOGY OF LIVING THINGS, THE PIS OPERATES TOTAL BODY ELECTRON, NUCLEON POPULATION FUNCTIONS, THE NANO UNITS AND MICRO UNIT TASK OPERATIONS, UNDER PIS – CLOUD ORDERS THROUGH PCS

38 – TOTAL BODY ELECTRONS, NEUTRONS, PROTON, NANO UNIT AND MICRO UNITS FUNCTION UNDER PIS

39 - SYNTHESIS OF HORMONES, PROTEINS, ENZYMES, GENES, CELLS, TISSUES TAKE PLACE UNDER PIS ORDERS

40 – CELL GENESIS, BIOMOLECULE AND GENES GENESIS UNDER PIS PARTICLE CLOUD HIERARCHY ORDERS

41 – PIS OF THE DOMINENT MOLECULES (SUCH AS HORMONES), PIS OF BIOMOLECULES, RELATIONSHIPS OF PIS OF DOMINENT MOLECULES, PIS OF DOMINENT BIOMOLEULES WITH PIS OF THE CNS CENTERS

42 – MOLECULAR EVOLUTION IS EQUAL TO LIVING THING GENESIS

43 - MOLECULAR EVOLUTION IN FUNDAMENTAL PARTICLES

44 – MOLECULAR EVOLUTION IN MICRO UNITS

45 - MOLECULAR EVOLUTION AND LIVING THING GENESIS

46 – AUTONOMOUS MAINTENANCES OF LIVING THINGS

47 – THE FUNDAMENTAL PARTICLE SOURCES FOR ELECTRON GENESIS, NEUTRON GENESIS, PROTON GENESIS, DOMINANT MOLECULES AND DOMINANT BIOMOLECULES

48 – MALFUNCTIONING PIS, AND NORMAL FUNCTIONING PIS

49 – NORMAL FP –CLOUDS, ABNORMAL FP –CLOUDS, NORMAL INTERNAL ATOM PARTICLE COMPOUND GENESIS, AND ABNORMAL PARTICLE COMPOUND GENESIS INSIDE ELECTRONS AND NUCLEONS

50 – DISEASE OF CNS – PIS AND CNS –PCS, CONTROL OF TOTAL BODY ATOM POPULATIONS BY CNS – PIS ATOMS

51 – ABNORMAL CONSTRUCTION NANO UNITS, SICK ATOMS AND NORMAL ATOMS

52 – THE INTERNAL ELECTRON NUCLEON DISEASES SIGNS AND SYMPTOMS ARE SIMILAR TO CELL DISEASES

53 – SIGNS OF SICK CARDIAC – PIS, SICK ATOMS DISORDERS SIGNS AND SYMPTOMS, SICK CARDIAC – PCS SYNDROMES, SICK CNS – PIS SYNDROMES, AND SICK CNS – PCS SYNDROMES, CARDIAC PARTICLE CLOUD DISORDERS

54 – SICK GIS – PIS DISORDERS, SICH GUS – PIS DISORDERS, ABNORMAL PCS DISORDERS OF GIS AND GUS, SICK ATOMS AND NORMAL ATOMS

55 - PSYCHIATRIC DISEASES ARE ABNORMAL PARTICLE CLOUD TRANSMISSION DISEASES, SICK CNS ATOMS PRODUCE ABNORMAL PSYCHE (SICK PSYCHE GENESIS SYNDROMES) NORMAL CNS ATOMS PRODUCE NORMAL PSYCHE

56 – ABNORMAL FUNDAMENTAL PARTICLE CLASSES

57 - PARTICLE CLOUD CIRCULATION SYSTEMS (PCS), INDIGENOUS – PCS, EXO. –PCS, PCS DISEASES

58 –COMMUNICATION OF TOTAL BODY ELECTRONS NUCLEONS ATOMS WITH EACH OTHER BY PCS, COMMUNICATION OF THE TOTAL BODY CELL POPULATION WITH EACH OTHER THROUGH PCS

59 – ABNORMAL CNS PARTICLE COMPOUND CONSTRUCTION OF ATOMS (SICK ELECTRONS NUCLEONS), ABNORMAL PCS AND PARTICLE CLOUDS, SIGNS AND SYMPTOMS OF THE SICK ATOMS, ELECTRONS, NUCLEONS.

60 – COLD FUNDAMENTAL PARTICLES

61 – ANGINA CRANIUM, ANGINA PEDIS, ANGINA TYMPANIC MEMBRANE

62 – PSYCHE GENESIS INSIDE CNS ELECTRONS, NEUTRON, PROTONS, INTELLIGENCE GENESIS AND ACCUMULATION OF KNOWLEDGE INSIDE CNS ELECTRONS, NEUTRONS, PROTONS,

MEMORIZATION PHENOMENON, PARTICLE CLOUD STORAGE AND RETRIEVAL PHENOMENONS

63 – UNIVERSE UNDER THE GRAVITON LAWS AND ORDERS, GRAVITON AND MOLECULAR EVOLUTION

64 – UNIT OF GRAVITON, GENESIS OF UNIVERSE UNDER GRAVITON FORCE, MOLECULAR EVOLUTION UNDER GRAVITON FORCE

65 – FUNDAMENTAL PARTICLE MOLECULE NEO GENESIS UNDER GRAVITON FORCES

66 – NANO UNIT NEO GENESIS UNDER GRAVITON FORCES, GRAVITON FORCES CREATED MOLECULAR EVOLUTION

67 – GENESIS OF LIVING THINGS BY GRAVITON FORCES, EVOLUTION IN BIOMOLECULES, CELLS, TISSUES, BODY ORGANS, AND SYSTEMS UNDER GRVITON FORCES REGULATIONS AND LAWS

68 – FUNDAMENTAL BIOLOGICAL SCIENCES, PHYSICS, AND CHEMISTRY SCIENCES IN PLANTS

69 – PLANTS ELECTRON, NUCLEON GENESIS, ATOM GENESIS, CELL GENESIS, PLANT COLOR GENESIS

70 – EARTHS LIGHT FUNDAMENTAL PARTICLES, COLOR OF PLANTS, FP – BIOCHEMISTRY

71 – PLANT GENESIS AND ANIMAL GENESIS

72 – AUTONOMOUS SEQUENTIAL STEM CELL MUTATIONS PHASES ALTERNATING WITH CELL MEDIA CHEMICAL FORMULARY CHANGES PHASES, CREATE NEW TISSUE GENESIS, BODY ORGAN GENESIS, NEO BODY SYSTEM GENESIS AND GENESIS OF NEW PLANT SPECIES, AND GENESIS OF NEW ANIMAL SPECIES, CREATIONS OF LIVING SPECIES

73 – ELECTRON NUCLEON AND ATOM MUTATIONS AND PLANT NEO GENESIS

74 – ATOM MUTATION CAUSES STEM CELL MUTATIONS AND AUTONOMOUS SEQUANTIAL CELL MUTATIONS, MOLECULAR EVOLUTION PHASES EQUALS TO EMBRYONIC MUTATIONS

75 – THE INTELLIGENT PLANTS AND NON INTELLIGENT PLANTS, S – PIS, C – PIS, M- PIS, PCS – CIRCUITS BETWEEN THE PARTICLE INTELLIGENCE SYSTEM CENTERS (PIS), PLANTS NANO NEUROLOGY

76 – PLANTS PCS – CIRCUITS BETWEEN S – PIS, C – PIS, M- PIS, ONE HIRARCHY PCS – CIRCUIT BETWEEN DIFFERENT PLANTS PIS CENTERS

77 – PARTICLE CLOUD CURRENTS IN PLANTS, PCS OF THE PLANTS

78 – PLANTS CLASSIFICATIONS THROUGH PIS – PCS

79 – PLANTS BRAIN (PIS) FUNCTIONS

80 – ONE FUNCTION FROM PLANT BRAIN (PIS) IS HAZARD PREVENTION, THAT MEANS PLANTS PIS TASK IS NOT LET ENVIRONMENT HURT PLANTS. ALSO THE PIS HAS OTHER BIOLOGICAL NEEDED FUNCTIONS

81 – BIOLOGICAL TASKS OF PLANT PIS IN EMBRYO IS EQUAL MOLECULE EVOLUTION PIS TASKS OF PAST

82 – PLANTS ALZHEIMER DISEASES, PHENOMENON OF IMMUNITY AND PLANT DEFENSIVE STSTEMS

83 - THE PLANTS WITH NO PIS

84 – EFFECTS OF KINDS OF FUNDAMENTAL PARTICLE MOLECULES THAT CONSTRUCT PARTICLE COMPOUND CONSTRUCTIONS OF THE ELCTRONS AND NUCLEONS IN PLANT'S IMMUNITY AND DEFENSE SYSTEMS

85 - HIERARCHY ORDER SYSTEMS BETWEEN: S – PIS, C – PIS, M – PIS, THOUGHT CURRENTS AND PSYCHE GENESIS IN PLANTS

86 – PSYCHE GENESIS AND THOUGHT CURRENT GENESIS IN PLANTS, KNOWLEDGE ACCUMULATION INSIDE THE PLANT ATOMS AND CELLS, AND PLANTS DECISION MAKING PHENOMENONS

87 - THE LIGHT LASER SUPERSONIC MICROWAVE FUNDAMENTAL PARTICLES WEAPONS INJURIES, AND CRIMES AGAINST HUMANITY

88 – LASER MICROWAVE PARTICLES SHOTS INTO BRAIN AND HEAD INJURIES AND DESTRUCTION OF INTERNAL ELECTRON, NEUTRON AND PROTON CONSTRUCTION INJURIES, SIGNS AND SYMPTOMS, AND NEO GENESIS OF THE PATHOLOGICAL CONSTRUCTED ATOMS

89 – PHYSIOLOGICAL FUNCTIONS OF THE CERUMEN

90 – WAR SYNDROMES, SYMPTOMS AND SIGNS OF CNS INTERNAL ELECTRON NUCLEON FUNDAMENTAL PARTICLE DESTRUCTIONS AND INJURIES BY POWERFUL LETHAL LIGHT LASER MICROWAVE PARTICLE FORCES

91 - LASER MICROWAVE GUN SHOT OF LUNGS AND HEART, HOW FUNDAMENTAL PARTICLES TRAVEL IN DIFFERENT TISSUES AND DAMAGING VITAL ORGANS PIS- PCS BIOLOGICAL CHEMICAL PHYSICAL FUNCTIONS

92 – THE WEAPONS THAT THEIR INJURIES NOT LEAVE ANY TRACE FROM CRIMES

93 – TREATMENTS OF LASER MICROWAVE INJURIES

SOME OTHER ABNORMAL FUNDAMENTAL PARTICLE CLOUD TRANSMISSION DISEASES

THE FUNDAMENTAL PARTICLES (F.P.) OR UNIT'S OF THE MATTER

THE EXISTING LAWS AND ORDERS OF MATTER IN UNIVERSE

GENERAL CHIMISTRY OF FUNDAMENTAL PARTICLES

ORGANIC CHEMISTRY, BIOLOGICAL CHEMISTRY OF FUNDAMENTAL PARTICLES

FUNDAMENTAL PARTICLES ARE UNITS OF THE MATTER, THE FUNDAMENTAL PARTICLES CONSTRUCT THE EXISTING DIFFERENT ELECTRONS, NEUTRONS, PROTONS, QUARKS PARTICLE COMPOUND CONSTRUCTIONS, ALL OVER UNIVERSE, FOR ALL PLANETS UNDER THE MATTER'S PHYSICAL, CHEMICAL, BIOLOGICAL (RELATIVELY FIX, PARTICLE SPECIFIC, PLANET SPECIFIC) RULES, LAWS AND ORDERS.

EACH GIVEN FUNDAMENTAL PARTICLE CLASS AT ANY GIVEN PLANET, AT ANY LOATION OF UNIVERSE, POSSESSES WELL DEFINED, RELATIVELY FIX PHYSICAL, CHEMICAL, BIOLOGICAL PARTICLE CLASS SPECIFIC PROPERTIES, THE FUNDAMENTAL PARTICLES OF DIFFERENT PLANETS AT DIFFERENT UNIVERSES, DIFFERENT GALAXIES, AND DIFFERENT PLANETS ALL ARE DIFFERENT FROM EACH OTHER.

AT EACH GIVEN PLANET THE EXISTING DIFFERENT FUNDAMENTAL PARTICLE CLASSES ARE WELL DEFINED, PLANET SPECIFIC INDIGENOUS FUNDAMENTAL PARTICLE CLASSES, AND THEIR CHEMICAL, BIOLOGICAL AND PHYSICAL PROPERTIES ARE PLANET SPECIFIC AT MOST.

IN EACH GIVEN PLANET, THE FUNDAMENTAL PARTICLE'S WEIGHT, GRAVITY, FREQUENCY, WAVE-LENGTH, POLE, SPEED, SPIN, TRANSIT ROUTES, TILTS, ROTATIONS, COLOR, SIZE, SHAPE, PARTICLE'S COLONY SHAPE, PARTICLE'S COLONY SIZE, PARTICLE'S SENSIBILITY, PARTICLE'S ENERGY KIND, PARTICLE ENERGY CONTENT, PARTICLE GRAVITON FORCES, AND PARTICLE'S MASS AND GRAVITON FORCE QUANTITIES, PATTERNS OF PRODUCED PARTICLE TRAVEL AND CURRENTS, PARTICLE CLOUDS, PCS, PIS, ETC., ARE ALL FUNDAMENTAL PARTICLE SPECIFIC, AND LOCATION SPECIFIC, THESE PARTICLE PROPERTIES VARIES IN DIFFERENT UNIVERSAL LOCATIONS FOR DIFFERENT GALAXIES, DIFFERENT PLANETS, AND DIFFERENT UNIVERSES.

THE FUNDAMENTAL PARTICLES OF DIFFERENT PLANETS ARE DIFFERENT FROM EACH OTHER, THE FUNDAMENTAL PARTICLES AT DIFFERENT PLANETS CONSTRUCT DIFFERENT KIND INTERNAL ELECTRON NUCLEON FUNDAMENTAL PARTICLE COMPOUNDS, DIFFERENT KIND ELECTRONS MOLECULAR CONSTRUCTIONS, DIFFERENT KIND NEUTRONS, DIFFERENT KIND PROTONS FUNDAMENTAL PARTICLE COMPOUND CONSTRUCTIONS, CONSTRUCT DIFFERENT KIND ATOMS, AND NANO UNITS. UNDER AUTONOMOUS SEQUENTIAL INTER FUNDAMENTAL

PARTICLE MOLECULAR CHEMICAL INTERACTION CYCLES AND CHAINS (A. S. I. F.P. Mol. C.I.C.).

THE DIFFERENT PLANETS AT DIFFERENT LOCATIONS OF THE UNIVERSE HAVE DIFFERENT KIND ELECTRONS GENESIS, DIFFERENT KIND NUCLEON GENESIS, AND DIFFERENT KIND ATOM GENESIS, AND NANO UNIT GENESIS, KINDS OF ATOMS AT BLACK HOLES ARE DIFFERENT CLASSES (BLACK MATTER), AND ATOMS CONSTRUCTIONS AT BIOFRIENDLY PLANETS SUCH AS EARTH ARE THE ELEMENTS AS WE SEE IN EARTH, AND SOME OF EARTH'S ELEMENTS ARE AT PERIODIC TABLE. SOME PLANET'S FUNDAMENTAL PARTICLES ARE BIOFRIENDLY AND SENSIBLE, IN MANY OTHER PLANET'S PARTICLES ACROSS UNIVERSE THE FUNDAMENTAL PARTICLES ARE BIO-LETHAL AND ARE NOT SENSIBLE.

FUNDAMENTAL PARTICLE COLONY (P. Cl.)

FUNDAMENTAL PARTICLE POPULATION OF A PARTICLE COLONY.

THE MOST FUNDAMENTAL PARTICLES CREATE WAVE SHAPE COLONIES, THE PARTICLE COLONIZATIONS IN WAVE SHAPES HELP THE FUNDAMENTAL PARTICLES TRAVEL AND TRANSIT FROM ONE LOCATION TO OTHER EASILY WITH HIGH SPEEDS. PARTICLES CAN TRAVEL IN SPACE, IN PERIPHERAL ATOM SPACES, OR INSIDE THE INTERNAL ELECTRON, NUCLEON PARTICLE CIRCULATION SYSTEMS, THE SPEEDS OF THE FUNDAMENTAL PARTICLES VARIES REMARKABLY ONE

FROM THE OTHER, IN DIFFERENT MEDIAS, IN DIFFERENT UNIVERSAL LOCATIONS, AND AT DIFFERENT PLANETS.

AT EARTH, THE POWERFUL X – FUNDAMENTAL PARTICLE COLONIES, ELECTRIC PARTICLE CURRENTS, OR GAMMA FUNDAMENTAL PARTICLE COLONIES TRAVEL AND THEIR PARTICLE CIRCULATION SYSTEMS EITHER IN EXTERNAL ATOM OR IN INTERNAL ATOM PARTICLE CURRENTS AND PARTICLE CIRCULATIONS SYSTEMS, ARE MUCH FASTER THAN THE EARTH'S WEAK ENERGY SONIC FUNDAMENTAL PARTICLE COLONIES, SONIC PARTICLE INTERNAL ELECTRON NUCLEON CIRCULATION SYSTEMS, AS WELL AS AT SONIC PARTICLE'S EXTERNAL ATOM PARTICLE CIRCULATION SYSTEMS. THE EARTH'S LIGHT PARTICLE COLONIES ARE FASTER THAN SONIC PARTICLE CURRENTS TRAVELS.

IN OTHER HIGH GRAVITON, HIGH WEIGHT, HIGH ENERGY VICIOUS BIO CIDAL PLANETS WITH HIGH ENERGY FUNDAMENTAL PARTICLES COLONIES AND HIGH SPEED PARTICLE CIRCULATION SYSTEMS, THEIR DIFFERENT HIGH ENERGY, POWERFUL BIO CIDAL FUNDAMENTAL PARTICLE COLONY AND CLOUDS CLASSES CIRCULATION SYSTEMS AND COLONIZATIONS, ARE MILLIONS FOLDS FASTER, VICIOUS AND POWERFUL THAN WEAK EARTH'S BIO FRIENDLY FUNDAMENTAL PARTICLE COLONIZATIONS AND PARTICLE CLOUDS.

THE FUNDAMENTAL PARTICLES ADDITIONALLY POSSESSES CAPABILITIES UNDER DIFFERENT PHYSICAL CHEMICAL CONDITIONS TRANSFORM INTO DIFFERENT OTHER FORMS OF PARTICLE COLONIZATIONS, DIFFERENT HOMOGENEOUS, OR NON- HOMOGENEOUS COLONY AND CLOUD PATTERNS.

FUNDAMENTAL PARTICLES OCCUPY SPACE IN DIFFERENT COLONIZATION PATTERNS, INDIVIDUAL PARTICLES TYPES, COMPOSITE TYPE PARTICLE COLONIZATIONS, STREPTO PATTERNS PARTICLE COLONIZATION, STAPHYLO TYPE FUNDAMENTAL PARTICLE COLONY, AND ENORMOUS OTHER ENORMOUS DIFFERENT PATTERNS FUNDAMENTAL PARTICLE COLONIZATIONS.

THE POPULATION OF THE FUNDAMENTAL PARTICLE - COLONY

THE TOTAL PARTICLE POPULATION IN ONE FUNDAMENTAL PARTICLE COLONY, EQUALS TO THE TOTAL FUNDAMENTAL PARTICLE POPULATION NUMBERS, WHICH EXIST IN ONE WAVE LENGTH. THE DIFFERENT FUNDAMENTAL PPARTICLE CLASSES HAVE DIFFERENT FUNDAMENTAL PARTICLE POPULATIONS, AND DIFFERENT WAVE LENGTH.

IN OTHER WORDS, THE TOTAL FUNDAMENTAL PARTICLE NUMBERS (OR PARTICLE POPULATIONS), THAT EXIST IN ONE WAVE – LENGTH, IS EQUAL TO ONE FUNDAMENTAL PARTICLE COLONY, FOR THAT GIVEN PARTICLE CLASS.

DIFFERENT FUNDAMENTAL PARTICLE CLASSES HAVE DIFFERENT PARTICLE POPULATION NUMBERS PER COLONY AND DIFFERENT PARTICLE COLONY. THE FUNDAMENTAL

PARTICLE COLONY SIZE AND TOTAL FUNDAMENTAL PARTICLE POPULATION NUMBERS AT EACH WAVE LENGTH OF A GIVEN FUNDAMENTAL PARTICLE CLASS ALWAYS ARE WELL DEFINED, FIXED AND PARTICLE SPECIFIC. AGAIN COLONIES OF DIFFERENT PARTICLES ARE DIFFERENT FROM EACH OTHER.

EARTH'S FUNDAMENTAL PARTICLES

FREE FUNDAMENTAL PARTICLE CLOUDS, PARTICLE COLONIES, AND PARTICLE COMPOSITES

IN UNIVERSE EACH GIVEN PLANET POSSESSES ITS OWN PLANET SPECIFIC FUNDAMENTAL PARTICLE CLASSES. THE EARTH'S FUNDAMENTAL PARTICLES MOSTLY ARE LESS WEIGHT (LESS FP MASS), LESS ENERGY, WEAK BIO FRIENDLY, LESS GRAVITON FORCE, DIFFERENT CLASSES OF BENIGN FUNDAMENTAL PARTICLES.

IN UNIVERSE, SOME FUNDAMENTAL PARTICLES ARE FREE, NOT PART OF ANY NANO UNITS CONSTRUCTION, PRODUCING ENORMOUS DIFFERENT KIND PARTICLE CLOUD CONSTRUCTIONS, DIFFERENT FORMS FUNDAMENTAL PARTICLE COMPOSITES, AND DIFFERENT PATTERNS PARTICLE COLONIES.

THESE FUNDAMENTAL PARTICLE CLOUDS, COLONIES AND COMPOSITES, EITHER LOCATED IN PERIPHERAL ATOM

SPACE (PAS) IN DIFFERENT UNIVERSAL LOCATION, OR ARE AIRBORNE AT DIFFERENT UNIVERSAL LOCATIONS, IN SPACE AS FREE PARTICLE CLOUDS, HAVE OWN PATTERNS OF PARTICLE CURRENTS, AND FUNDAMENTAL PARTICLE CLOUD CIRCULATION SYSTEMS (PCS), AND OTHER FORMS.

FUNDAMENTAL PARTICLES ALSO UNDER AUTONOMOUS SEQUENTIAL INTER FUNDAMENTAL PARTICLE CHEMICAL INTERACTION CYCLES AND CHAINS (A. S. I. FP. Mol. C. I. C.) INSIDE ELECTRONS NUCLEONS SUBSYSTEM CHEMICAL LAB.S, OR IN OUT OF NANO UNIT LAB.S OF UNIVERSE, CHEMICALLY INTERACT WITH EACH OTHER, PRODUCE NEW PARTICLE COMPOUNDS, AND CONSTRUCT INTERNAL ATOM, INTERNAL ELECTRON, INTERNAL NUCLEONS PARTICLE MOLECULAR COMPOUND CONSTRUCTIONS OF LIVE OR NON LIVE MOLECULAR COMPOUNDS IN ALL DIFFERENT PLANETS, INCLUDING CONSTRUCTIONS OF THE EARTH'S ELEMENTS.

IN EARTH, ONLY A FEW FUNDAMENTAL PARTICLE CLASSES WHICH CONSTRUCTING LIVING THINGS ELECTRONS, PROTONS, NEUTRONS, QUARKS, AND NANO UNITS ARE SENSIBLE. BUT THE LARGE NUMBER OTHER REMAING DIFFERENT FUNDAMENTAL PARTICLE CLASSES OF EARTH, ALMOST ENTIRELY ARE NON SENSIBLE, AND NOT DISCOVERED YET.

THE EARTHS FUNDAMENTAL PARTICLES DIVIDE INTO THREE MAJOR CLASSES:

1 – BIO HOSTILE FUNDAMENTAL PARTICLE CLASS.

2 – BIO FRIENDLY FUNDAMENTAL PARTICLE CLASS.

3 – INTERMEDIATE CLASSES OF FUNDAMENTAL PARTICLES.

EARTH'S BIO FRIENDLY SENSIBLE PARTICLES

VISIBLE BIO FRIENDLY LIGHT FUNDAMENTAL PARTICLE CLOUDS AND PARTICLE COLONIES

AUDIBLE SONIC FUNDAMENTAL PARTICLE COLONIES AND PARTICLE CLOUDS

BIO FRIENDLY SENSIBLE THERMAL, ELECTRIC FUNDAMENTAL PARTICLE CLASSES AND PARTICLE CLOUDS

EARTH'S BIO FRIENDLY FUNDAMENTAL PARTICLES CLASSIFY INTO TWO MAJOR CLASSES:

1 - NON SENSIBLE BIO FRIENDLY FUNDAMENTAL PARTICLE CLASSES, AT PRESENT TIME MOST OF THESE FUNDAMENTAL PARTICLE CLASSES HAVE NOT BEEN DISCOVERED, AND ARE LARGER THAN SENSIBLE PARTICLE CLASSES. THESE FUNDAMENTAL PARTICLES CONSTRUCT MOST OF PLANET EARTH'S INTERNAL ELECTRONS, NUCLEON PARTICLE

COMPOUND CONSTRUCTIONS. FOR BOTH OF LIVE AND NON LIVE NANO UNITS.

2 – DIFFERENT CLASSES OF WEAK, BIO FRIENDLY, SENSIBLE FUNDAMENTAL PARTICLES AND PARTICLE CLOUDS, ALSO ENTER INTO INTERNAL ELECTRON AND NUCLEON SUBSYSTEM UNITS CHEMICAL LAB. UNDER A. S. I. FP. Mol. CIC. CONSTRUCT INTERNAL ELECTRON, NUCLEON PARTICLE CLOUD COMPOUND CONSTRUCTIONS, FOR BOTH LIVING THINGS AND NON LIVING ELECTRONS, NEUTRONS, PROTONS, AND ATOMS.

EXAMPLES FROM, EARTH'S BIO FRIENSLY PARTICLE CLASSES ARE: 1 - SENSIBLE AND NON SENSIBLE ELECTRIC FUNDAMENTAL PARTICLE CLASSES. 2 – SENSIBLE AND NON SENSIBLE LIGHT FUNDAMENTAL PARTICLE CLASSES. 3 – BIO FRIENDLY NON SENSIBLE AND SENSIBLE THERMAL FUNDAMENTAL PARTICLE CLASSES. 4 – BIO FRIENDLY SENSIBLE AND NON SENSIBLE SONIC PARTICLE CLASSES. 5 - AND ETC.

ALL OF THESE BIO FRIENDLY SENSIBLE AND NON SENSIBLE FUNDAMENTAL PARTICLE CLASSES ALTOGETHER CONSTRUCT PARTICLE MOLECULAR COMPOUND CONSTRUCTIONS OF THE PLANET EARTH ATOMS, NUCLEONS, AND ELECTRONS. THESE ARE ALSO IN CHARGES OF OPERATIONS OF LIVING THINGS PHYSICAL, CHEMICAL BIOLOGICAL FUNCTIONS AND OPERATIONS.

EARTH'S SENSIBLE BIO FRIENDLY FUNDAMENTAL PARTICLE CLASSES

THE BIO FRIENDLY LIGHT FUNDAMENTAL PARTICLES (Y – FP) HAVE BEEN USED IN CONSTRUCTIONS OF THE EARTH'S LIVE AND NON LIVE INTERNAL NANO UNIT PARTICLE MOLECULAR COMPOUND CONSTRUCTIONS. THE INDIGENOUS EXISTING EARTH'S FUNDAMENTAL PARTICLES, SUCH AS LIGHT PARTICLES, SONIC PARTICLES, THERMAL AND ELECTRIC FUNDAMENTAL PARTICLES, ETC. TRAVEL AIRBORNE, OR TRANSIT THROUGH PERIPHERAL ATOMS SPACE PARTICLE CIRCULATION SYSTEMS (PAS – PCS), AND FROM PAS PARTICLE CURRENTS ENTER INTO INTERNAL ELECTRON, NUCLEON PCS.

THE BIO FRIENDLY PARTICLE CURRENTS, THROUGH INTERNAL PERIPHERON, NUCLEUS PCS TRAVEL INTO ELECTRONS, NUCLEON'S SUBSYSTEM UNITS CHEMICAL LAB.S. UNDER A. S. I. FP. Mol. CIC COMBINE WITH INTERNAL ELECTRON, NUCLEON PARTICLE COMPOUNDS, PRODUCE NEW PARTICLE COMPOUND CONSTRUCTIONS INSIDE NANO UNITS. AND CONSTRUCT NEW INTERNAL ELECTRONS, NUCLEONS PARTICLE MOLECULAR COMPOUND CONSTRUCTIONS OF THE EARTH'S EXISTING ELEMENTS (ATOMS).

THIS IS PHENOMENON OF ATOM GENESIS, ELECTRON GENESIS, AND NUCLEON GENESIS, UNDER PATHS AND ORDERS OF MOLECULAR FUNDAMENTAL PARTICLE EVOLUTION.

IN PRESENT TIME, THE ONGOING A. S. I. FP. Mol. CIC. ARE COPIES OF PAST WORLD HISTORY, PARTICLE MOLECULAR EVOLUTION PATHS AND ORDERS. PRODUCTION OF NEW PROTONS, NEUTRONS, ELECTRONS UNDER PATHWAYS OF MOLECULAR EVOLUTIONS THROUGH REGENERATIVE AND DEGENERATIVE A. S. I. FP. Mol. CIC. STILL REPEATING EXACTLY THE SAME ORDERS AND PATHS OF MOLECULAR EVOLUTIONS AFTER MILLIONS YEARS EXACTLY THE SAME.

THE DIFFERENT KINDS LIGHT FUNDAMENTAL PARTICLE COLONIES, THE LIGHT FUNDAMENTAL PARTICLE CLOUDS (Y. FP. – Cl.), THE DIFFERENT KIND BIO FRIENDLY LIGHT PARTICLES (Y – FP), AND DIFFERENT KINDS Y. S. T. E. – FP – I.I. – P. Cl. ALL TODAY HAVE BEEN USED IN CONSTRUCTION OF THE LIVE, AND NON LIVE MOLECULAR COMPOUND CONSTRUCTION OF THE PLANET EARTH'S INTERNAL ELECTRONS, NEUTRONS, PROTON'S MOLECULAR COMPOUND CONSTRUCTIONS, THE ELECTRONS, NUCLEONS HAVE BEEN CONSTRUCTED THROUGH THE USE OF THESE PARTICLE CLOUD MOLECULAR STRUCTURES AT PLANET EARTH TODAY EQUALS TO EVOLUTION.

THE LIGHT FUNDAMENTAL PARTICLES (Y – FP), THE LIGHT FUNDAMENTAL PARTICLE CLOUDS (Y. FP. – Cl.), LIGHT PARTICLE CLOUD COLONIES, LIGHT FUNDAMENTAL PARTICLES INFORMATION, IMAGE PARTICLE CLOUDS (Y. – FP – I.I. – P. Cl.) ENTER INTO PLANET EARTH'S ELECTRONS,

NUCLEONS, AND COMBINE WITH ATOMS DIFFERENT INTERNAL ELECTRON, NUCLEON PARTICLE MOLECULAR COMPOUNDS, AND CONSTRUCT THE EARTH'S ELECTRONS, NUCLEONS PARTICLE COMPOUND MOLECULAR STRUCTURES.

THE EXOGENOUS S. Y. T. E. – FP – I.I. – P. Cl. AND OTHER FUNDAMENTAL PARTICLE CLOUDS AIRBORNE ENTER INTO PCS OF THE DIFFERENT ATOMS INTERTNAL ELECTRONS NUCLEONS PCS, TRAVEL INSIDE DIFFERENT INTERNAL ELECTRONS AND NUCLEONS SUBSYSTEMS AND SYSTEM UNITS, SELECTIVELY UNDER REGENERATIVE A. S. I. FP Mol. CIC COMBINE WITH INTERNAL ELECTRON NUCLEON PARTICLE COMPOUNDS, AND CONSTRUCT NEW LIGHT CLOUD PARTICLE MOLECULAR COMPOUND ELECTRONS, PROTONS, NEUTRONS, AND OTHER INTERNAL ATOM NANO UNITS CONSTRUCTIONS. THIS IS PHENOMENON OF NEO PROTON GENESIS, NEO NEUTRON GENESIS, NEO ELECTRON GENESIS, AND NANO UNIT GENESIS.

THE DIFFERENT EXOGENOUS LIGHT FUNDAMENTAL PARTICLES, AND FUNDAMENTAL PARTICLE CLOUDS CIRCULATE BETWEEN DIFFERENT ATOMS OF DIFFERENT ENVIRONMENTAL SUBJECTS THROUGH EXOGENOUS PARTICLE CIRCULATION SYSTEMS (EX. – PCS) CURRENTS.

THE RECIPIENT ELECTRONS, NUCLEONS ATTRACTIVE GRAVITON FORCES, ATTRACT ENVIRONMENTAL FUNDAMENTAL PARTICLES INFORMATION, IMAGE PARTICLE CLOUDS, FINALLY EXOGENOUS FUNDAMENTAL PARTICLES ENTER INTO DIFFERENT INTERNAL ELECTRON, AND NUCLEON PCS OF THE RECIPIENT SUBJECTS CNS ATOMS.

STORAGE OF ENVIRONMENTAL EVENTS INFORMATIONS AND IMAGES IN PARTICLE CLOUD FORMS (Y. S. T. E. – FP- I.I. –P. Cl.), INSIDE CNS ELECTRONS, NUCLEONS IN FUNDAMENTAL PARTICLE INFORMATION AND IMAGE PARTICLE CLOUD COMPOUND FORMS, TAKE PLACE UNDER REGENERATIVE AND DEGENERATIVE A. S. I. FP. Mol. CIC.

IN LIVING THINGS, CONTINUOUS INTERACTIONS OF THESE PARTICLE CLOUDS WITH EACH OTHER, ALSO THEIR INTERACTIONS WITH OTHER INCOMING EXOGENOUS PARTICLE CLOUDS, AND Y. S. T. E. – FP – I.I. – P. Cl., ALSO THEIR INTERACTIONS WITH OTHER PRE-EXISTING CNS INTERNAL ELECTRON, NUCLEON PARTICLE CLOUD COMPOUNDS OF DIFFERENT CNS CENTERS ELECTRONS AND NUCLEONS POPULATIONS, UNDER A. S. I. FP. Mol. CIC. BIOLOGICALLY SENSE AS PSYCHE AND THOUGHT CURRENTS.

SENSING PARTICLE CLOUD INTERACTIONS DURING SLEEP CREATE DREAMS. THESE FUNDAMENTAL PARTICLE INFORMATION AND IMAGE PARTICLE CLOUD COMPOUND CONSTRUCTIONS ALSO ARE PHENOMENONS OF INTERNAL ELECTRON NUCLEONS INTELLIGENCE AND KNOWLEDGE STORAGE SYSTEMS, AND KNOWLEDGE GENESIS THROUGH PARTICLE CLOUD COMPOUND CONSTRUCTIONS. THESE ARE STORAGES OF KNOWLEDGE, INFORMATION AND IMAGES, SCIENCE INSIDE CNS ELECTRONS AND NUCLEONS.

THESE ARE STORAGE OF OUTSIDE WORLD INFORMATIONS AND IMAGES, ACCUMULATE IN PARTICLE CLOUD COMPOUND FORMS, INSIDE CNS ATOMS, ELECTRONS, NUCLEONS. AND HAS EXTREME IMPORTANCE IN LEARNING, TEACHING, TRAINING, FROM BIRTH TO ALL LIFE LONG, USED

IN SCHOOLS, HOME, COLLEGES, UNIVERSITIES, AND STORED PARTICLE CLOUDS CAN BE CHANGED, OR ERASED, RECALLED (RETRIEVED) UNDER REVERSE DEGENERATIVE A. S. I. FP. Mol. CIC, ETC.

ABOVE IS PHENOMENON OF PSYCHE GENESIS, AND CIRCULATIONS OF THE PARTICLE CLOUDS BETWEEN DIFFERENT CNS ELECTRONS, NUCLEONS DIFFERENT CNS CENTERS ELECTRONS, NUCLEONS PRODUCE THOUGHT CURRENTS, AND SENSE AS THOUGHT CURRENTS AND PSYCHE BIOLOGICALLY.

CLASSIFICATION OF SENSIBLE LIGHT PARTICLES

THE LARGE NUMBERS OF FUNDAMENTAL PARTICLE CLASSES IN EARTH ARE BIO FRIENDLY, NON DISCOVERED NON SENSIBLE OR SENSIBLE FUNDAMENTAL PARTICLE CLASSES, WHICH THESE PARTICLES CONSTRUCT MOST OF INTERNAL ELECTRON NUCLEON FUNDAMENTAL PARTICLE COMPOUND MOLECULAR CONSTRUCTIONS OF THE EARTHS INTER ELECTRON, NUCLEON CONSTRUCTIONS.

LIGHT FUNDAMENTAL PARTICLES OF PLANET EARTH, CAN BE CLASSIFIED UNDER LIGHT FUNDAMENTAL PARTICLE'S COLOR DIFFERENCES, OR LIGHT PARTICLE'S QUANTUM ENERGY CONTENTS, ENERGY KIND DIFFERENCES, ALSO LIGHT PARTICLES CAN BE DIFFERENTIATED UNDER THEIR SENSIBILITY, OR NON SENSIBILITIES, PARTICLE SHAPE, SIZE,

WEIGHT, AND PARTICLE GRAVITON FORCE QUANTITIES, CLASSIFICATIONS OF PARTICLES THROUGH FUNDAMENTAL PARTICLE COLONY SHAPES AND FUNDAMENTAL PARTICLE COLONY POPULATION SIZES, WAVE PATTERNS AND SIZES, ETC.

THE ABOVE FUNDAMENTAL PARTICLE PROPERTIES ARE PART OF PERMANENT INDIGENOUS, INHERITANCE FUNDAMENTAL PARTICLE CHARACTERS, AND DIFFERENTIATES ONE PARTICLE FROM OTHER.

THE CLASSIFICATION OVER COLOR AND SENSIBILITY:

1 - RED COLOR LIGHT FUNDAMENTAL PARTICLE (OR THERMAL FUNDAMENTAL PARTICLE) CLASSES.

THIS CLASS PARTICLES CAN PROVIDE QUANTUM LIGHT ENERGY AND THERMAL ENERGY TO QUANTUM LOCATIONS, IN ORDER TO ACHIEVE QUANTUM TASKS, WHO ARE IN NEED OF THERMAL OR LIGHT ENERGIES.

2 - YELLOW COLOR LIGHT FUNDAMENTAL PARTICLE CLASS.

3 - GREEN COLOR LIGHT FUNDAMENTAL PARTICLE CLASSES (THE LIGHT ENERGY PROVIDER PARTICLES FOR NANO LOCATIONS, TO ACHIEVE NANO TASKS).

4 - BLUE COLOR LIGHT AND ELECTRIC FUNDAMENTAL PARTICLE CLASSES.

THE ELECTRIC ENERGY PROVIDERS AND LIGHT ENERGY DONORS FOR NON LIVE OR LIVE QUANTUM LOCATIONS, TO ACHIEVE NEEDED WORKS.

5 - VIOLET COLOR LIGHT AND ELECTRIC FUNDAMENTAL PARTICLE CLASSES.

QUANTUM LIGHT AND ELECTRIC ENERGY DONOR FOR BIOLOGICAL NANO SITE, OR OTHER NON LIVE OR LIVE LOCATIONS, TO ACHIEVE QUANTUM TASKS, THROUGH PROVIDING LIGHT AND ELECTRIC ENERGY.

6 - WHITE COLOR COMPOSITE LIGHT FUNDAMENTAL PARTICLES.

ALL OF ABOVE FUNDAMENTAL PARTICLES ALSO PROVIDE INDIVIDUAL FUNDAMENTAL PARTICLES TO ANY QUANTUM LOCATION TO BE USED IN PARTICLE MOLECULAR CONSTRUCTIONS, THROUGH A. S. I. FP. Mol. CIC, AND CONSTRUCT MOLECULAR CONSTRUCTIONS OF DIFFERENT ELECTRONS, NEUTRONS, PROTONS AND ATOMS.

ADDITIONALLY, THE LIGHT FUNDAMENTAL PARTICLES CAN BE CLASSIFIED, UNDER THE BASIS IF THEY ARE BIO HOSTILE, OR BIO FRIENDLY.

THE ULTRAVIOLET ELECTRIC FUNDAMENTAL PARTICLE CLASS: ULTRAVIOLET IS THE WEAKEST ELECTRIC FUNDAMENTAL PARTICLE IN EARTH. THE ULTRAVIOLET PARTICLE'S QUANTUM ELECTRIC ENERGY CONTENT IN ONE ULTRAVIOLET FUNDAMENTAL PARTICLE IS EQUAL TO ONE QUANTUM UNIT, ELECTRIC ENERGY.

THE BIO HOSTILE OR BIO FRIENDLY LIGHT FUNDAMENTAL PARTICLE CLASSES

MOST OF THE NON DISCOVERED BIO FRIENDLY LIGHT, THERMAL, ELECTRIC, SOUND FUNDAMENTAL PARTICLE CLASSES TOGETHER WITH SOME OF KNOWN OR NON KNOWN BIO HOSTILE PARTICLE CLASSES HAVE BEEN USED IN EARTH'S ELEMENTS TO PRODUCE THE FUNDAMENTAL PARTICLE CLOUD MOLECULAR COMPOUND CONSTRUCTIONS OF THE EARTH'S DIFFERENT ELECTRONS, NEUTRONS, PROTONS, NANO UNITS, ELEMENTS PARTICLE COMPOUND AND FUNDAMENTAL PARTICLE COMPOSITE CONSTRUCTIONS.
IN CONSIDERATION OF LIGHT PARTICLE CLASS, THE LIGHT ENERGY QUANTUM CONTENT OF THE PARTICLE, THE KIND OF LIGHT PARTICLE CLASS, AND LIGHT ENERGY QUANTITIES IN BIO HOSTILE AS WELL AS BIO FRIENDKY LIGHT FUNDAMENTAL PARTICLE'S DIFFERENT CLASSES ARE REMARKABLY DIFFERENT FROM EACH OTHERS, SOME TIMES DEPENDING TO PLANET THE DIFFERENCES RANGES ARE FROM THOUSANDS TO MILLIONS FOLDS DIFFERENT EITHER WEAK OR POWERFUL, VICIOUS IN COMPARING ONE TO THE OTHER. DEPENDING THE PLANET'S LIGHT PARTILES EITHER BELONGS TO A GIVEN POWERFUL BLACK WOLE PLANET OR ONLY IT BELONGS TO WEAK PLANET SUCH AS THE EARTH.

EACH GIVEN CLASSES OF ABOVE MENTIONED DIFFERENT COLOR BIO FRIENDLY LIGHT FUNDAMENTAL PARTICLES, IN EACH GIVEN CLASS POSSESSES DIFFERENT SIZE WAVE LENGTHS, DIFFERENT PARTICLE FREQUENCIES, DIFFERENT PARTICLE ENERGY KINDS, DIFFERENT QUANTITY FUNDAMENTAL PARTICLE ENERGY CONTENTS, DIFFERENT PARTICLE SPECIFIC FUNDAMENTAL PARTICLE WEIGHT AND PARTICLE GRAVITON FORCES, DIFFERENT KIND PARTICLE COLONY SIZES, PARTICLE COLONY SHAPES, ETC.

BUT IN EACH GIVEN LIGHT, ELECTRIC, THERMAL FUNDAMENTAL PARTICLE CLASS, THE ALL EXISTING PARTICLE SPECIFIC PHYSICAL, CHEMICAL, BIOLOGICAL PROPERTIES FOR EACH GIVEN CLASS, ALWAYS ARE WELL DEFINED, CERTAIN, CONSTANT, FIX, AND FUNDAMENTAL PARTICLE CLASS SPECIFIC PROPERTIES.

INFRARED FUNDAMENTAL PARTICLE CLASS

INFRARED THERMAL PARTICLES OR INFRARED LIGHT FUNDAMENTAL PARTICLES

$$(Y. Ir. - FP = T. Ir. - FP)$$

THE INFRARED FUNDAMENTAL PARTICLE POSSESSES TWO DIFFERENT KIND QUANTUM ENERGY QUANTITY IN ANY GIVEN ONE PARTICLE CONTENT, THE QUANTUM THERMAL

ENERGY AND THE QUANTUM LIGHT ENERGY IN THE CONTENTS OF ANY ONE SINGLE INFRARED PARTICLE.

THE FUNDAMENTAL PARTICLE'S QUANTUM ENERGY QUANTITY IN ONE PARTICLE ALSO PARTICLE'S ENERGY KIND, THE QUANTUM WEIGHT QUANTITY IN ONE PARTICLE, THE PARTICLE'S SIZE, SHAPE, GRAVITON FORCE, COLOR, COLONY SIZE, PARTICLE COLONY SHAPE, THE TOTAL PARTICLE POPULATION IN ONE COLONY, ETC. ARE INDIGENOUS PROPERTIES OF ANY GIVEN PARTICLE, THESE ARE ALL INHERITANCE CHARACTERISTICS FOR ANY GIVEN FUNDAMENTAL PARTICLE CLASS. THE INFRARED FUNDAMENTAL PARTICLE PRESENTLY POSSIBLE CLASSIFIED UNDER LIGHT FUNDAMENTAL PARTICLE CLASSES.

THE INFRARED FUNDAMENTAL PARTICLES CAN TRAVEL AIRBORNE, OR TRANSIT THROUGH PAS IN DIFFERENT LIQUID, SOLID, PLASMA MEDIAS ACCORDING PARTICLES ABILITY. THE INFRARED PARTICLES FROM PERIPHERAL ATOM SAPCES (PAS), ENTER INTO ATOMS, AND THROUGH ATOM'S PCS TRAVEL INSIDE INTERNAL ELECTRONS NUCLEON PCS, FINALLY ENTER INTO ELECTRONS, NUCLEONS SUBSYSTEM UNITS CHEMICAL LAB.S.

THE INFRARED FUNDAMENTAL PARTICLES UNDER REGENERATIVE A. S. I. FP. Mol. CIC. COMBINE WITH INTERNAL ELECTRON NUCLEON PARTICLE COMPOUNDS, PRODUCE INFRARED MOLECULAR COMPOUNDS INSIDE THE ELECTRONS AND NUCLEONS SUBSYSTEMS AND SYSTEM UNITS. AND BECOME PART OF PARTICLE MOLECULAR COMPOUNDS CONSTRUCTIONS OF ATOMS. THIS IS PHENOMENON OF INFRARED PARTICLE STORAGE SYSTEMS,

AND INFRARED THERMAL ENERGY STORAGE AS WELL, AND IT IS PHENOMENON OF INFRARED FUNDAMENTAL PARTICLES INFORMATION IMAGE PARTICLE CLOUD STORAGE PROCESS ALSO.

THE STORED INFRARED PARTICLE MOLECULAR COMPOUNDS, UNDER DEGENERATIVE A. S. I. FP. Mol. CIC. (WHICH ARE IN REVERSE CHEMICAL INTERACTION DIRECTIONS OF THE REGENERATIVE A. S. I. FP. Mol. CIC.) CAN BE FREED AND RELEASED FROM COMPOUND FORMS, INTO FREE INFRARED FUNDAMENTAL PARTICLE FORMS, WHICH IT IS THERMAL ENERGY PROVIDER, AND LIGHT ENERGY DONOR INTO NANO-SITES.

INFRARED PARTICLES CAN DELIVER THERMAL ENERGY TO TRIGGER START A. S. I. FP. Mol. CIC IN ANY GIVEN DIFFERENT NANO LOCATIONS, IT IS BIO FRIENDLY PARTICLE, AND INFRARED HAS REMARKABLE IMPORTANCES FOR ACHIEVING, STARTING, OR STOPPING, AND CONTROLS OF THE LIVING THING INTERNAL ELECTRON, INTERNAL NUCLEON CHEMICAL INTERACTIONS, INFRARED WITH ASSOCIATIONS OF THE OTHER BIO FRIENDLY THERMAL PARTICLES, ALTOGETHER THEY OPERATE THE LIVING THINGS INTERNAL ELECTRONS, NUCLEONS, ATOMS A. S. I. FP. Mol. CIC.

INFRARED FUNDAMENTAL PARTICLE STORE INSIDE NANO UNITS, AS THERMAL ENERGY DONORS AND LIGHT ENERGY PROVIDERS ARE CAPABLE TO OPERATE DIFFERENT KINDS PHYSICAL, CHEMICAL, BIOLOGICAL TASKS IN DIFFERENT NANO LOCATIONS, THROUGH PROVIDING QUANTUM

THERMAL ENERGIES OR QUANTUM LIGHT ENERGIES TO ANY NANO LOCATION.

IN LIVING THINGS, THE REGENERATIVE AND DEGENERATIVE A. S. I. FP. Mol. C.I.C. ALL ARE IN NEEDS OF INFRAREDS OR OTHER THERMAL PARTICLE'S DIFFERENT TYPES THERMAL ENERGIES, IN ORDER TO TRIGGER START A. S. I. FP. Mol. CIC. AND PRODUCE DIFFERENT NEEDED INTERNAL ELECTRON NUCLEON PARTICLE COMPOUND CONSTRUCTIONS.

IN GENERAL, THE DIFFERENT KIND FUNDAMENTAL PARTICLES DELIVER DIFFERENT KIND ENERGIES TO DIFFERENT KIND QUANTUM LOCATIONS, AND OPERATE DIFFERENT INTERNAL ELECTRON, INTERNAL NUCLEON, AND INTERNAL ATOMS PHYSICAL, CHEMICAL, BIOLOGICAL FUNCTIONS AND OPERATIONS.

ALL CHEMICAL INTERACTIONS IN ANY GIVEN QUANTUM BIOLOGICAL LOCATIONS, NEEDS SPECIFIC TEMPERATURES TO START, TO CONTINUE, PROCEED THE INTERACTIONS AND COMPLETE NANO TASKS INSIDE ELECTRONS, NUCLEONS SUBSYSTEM UNITS CHEMICAL LAB.S, THE NEEDED TEMPERATURE FOR ACHIEVING CHEMICAL AND BIOLOGICAL INTERACTION IN ANY NANO LOCATION ALL DELIVERED BY DIFFERENT ENERGY PROVIDER NANO UNITS AND FUNDAMENTAL PARTICLES.

ULTRAVIOLET FUNDAMENTAL PARTICLES

ULTRAVIOLET LIGHT FUNDAMENTAL PARTICLE OR ULTRAVIOLET ELECTRIC FUNDAMENTAL PARTICLE

(U v – Y. FP = U v – E. FP)

ULTRAVIOLET ELECTRIC PARTICLE IS WEAKEST ELECTRIC FUNDAMENTAL PARTICLE CLASS IN EARTH, ULTRAVIOLET BELONG TO ELECTRIC FUNDAMENTAL PARTICLE CLASS OF EARTH, BUT PRESENTLY IT HAS BEEN CLASSIFIED UNDER LIGHT FUNDAMENTAL PARTICLE CLASS. ULTRAVIOLET POSSESSES BOTH QUANTUM LIGHT ENERGY CONTENT, AS WELL AS ULTRAVIOLET ALSO POSSESSES QUANTUM ELECTRIC ENERGY CONTENT.

THE ULTRAVIOLET'S QUANTUM ELECTRIC ENERGY QUANTITY CONTENT, AT EACH GIVEN ONE SINGLE ULTRAVIOLET FUNDAMENTAL PARTICLE, IS LESS THAN QUANTUM ELECTRIC ENEGRY CONTENTS OF ALL OF THE OTHER INDUSTRIAL ELECTRIC FUNDAMENTAL PARTICLE CLASSES ELECTRIC – ENERGY – CONTENTS.

THE INDUSTRIAL ELECTRIC FUNDAMENTAL PARTICLES ARE BIO LETHAL. AND THEIR ELECTRIC ENERGY CONTENTS ARE MUCH LARGER QUANTITIES THAN ULTRAVIOLET ELECTRIC ENERGY CONTENTS PER ONE FUNDAMENTAL PARTICLE.

THE INDUSTRIAL ELECTRIC FUNDAMENTAL PARTICLE CONSTRUCTED ELECTRONS, NUCLEONS, ATOMS, ELECTRIC

ENERGY STORAGE QUANTITIES AND RETRIEVAL OF THESE ELECTRIC ENERGIES TO UT OFF ELECTRIC PARTICLE CONSTRUCTED ATOMS, ELECTRONS, NUCLEONS PRODUCE HIGH ELECTRIC ENERGY POWERFUL ELECTRIC CURRENTS, WHICH THESE ELECTRIC FUNDAMENTAL PARTICLE CURRENTS ARE USAGES IN COMMERCIAL INDUSTRIES. ALL ARE BIO LETHAL. NOT HAVE BIO FRIENDLY APPLICATIONS, EXCEPT IN DIAGNOSTIC AND THERAPEUTIC TOOLS.

THE ELECTRIC ENERGY CONTENT OF PARTICLES OF THE VICIOUS PLANETS SUCH AS BLACK HOLES ARE MILLIONS FOLDS MORE POWERFUL THAN THE EARTH'S INDUSTRIAL ELECTRIC FUNDAMENTAL PARTICLES.

THE INDUSTRIAL ELECTRIC FUNDAMENTAL PARTICLES UNDER REGENERATIVE A. S. I. FP. Mol. CIC. ARE USED IN PARTICLE COMPOUND MOLECULAR CONSTRUCTIONS OF NON LIVE ELECTRIC ATOMS, ELECTRONS, NUCLEON'S PARTICLE COMPOUND CONSTRUCTIONS

ABOVE ARE STORAGE OF INDUSTRIAL ELECTRIC PARTICLES WITH INDISTRIAL ELECTRIC ENERGY CONTENTS, INSIDE INDUSTRIAL ELECTRIC ATOMS, INDUSTRIAL HIGH POWER ELECTRIC ELECTRONS, AND HIGH ELECTRIC ENERGY INDUSTRIAL ELECTRIC NUCLEONS. ALL STORED IN THE FORMS HIGH ENERGY INDUSTRIAL ELECTRIC FUNDAMENTAL PARTICLE MOLECULAR COMPOUND CONSTRUCTIONS FORMS. THE RETRIEVAL OF THESE ELECTRIC PARTICLES THROUGH DEGENERATIVE A. S. I. FP. Mol. CIC. WILL PRODUCE FREE ELECTRIC FLOWING INDUSTRIAL PARTICLE CURRENTS, WITH BIO LETHAL CHARACTERISTICS. THESE ARE NOT WEAK BIO FRIENDLY ELECTRIC FUNDAMENTAL

PARTICLE MOLECULAR CONSTRUCTIONS, WHICH ARE USED IN LIVING THINGS BIO FRIENDLY ATOM CONSTRUCTIONS.

THE INDUSTRIAL ELECTRIC FUNDAMENTAL PARTICLE COMPOUNDS CONSTRUCT INDUSTRIAL ELECTRIC ELECTRONS AND NUCLEONS, AND UNDER DEGENERATIVE A. S. I. FP. Mol. CIC, ELECTRIC PARTICLES GET FREED AGAIN AS ELECTRIC PARTICLE CURRENTS.

UNIT OF QUANTUM LEVEL ELECTRIC ENERGY

ONE ULTRAVIOLET FUNDAMENTAL PARTICLE'S QUANTUM ELECTRIC ENERGY CONTENT IS EQUAL TO ONE-UNIT QUANTUM ELECTRIC ENERGY. FOR MEASUREMENTS OF THE QUANTUM LEVEL DIFFERENT ELECTRIC ENERGIES. IN OTHER WORDS, QUANTUM QUANTITY ELECTRIC ENERGY AMOUNT IN ONE EARHT'S ULTRAVIOLET FUNDAMENTAL PARTICLE, IS EQUAL TO ONE QUANTUM ELECTRIC ENERGY UNIT.

THE ULTRAVIOLET PARTICLES AT NON LIVE SUBJECTS ARE ELECTRIC ENERGY PROVIDERS FOR DIFFERENT QUANTUM SITES INSIDE ELECTRONS NUCLEONS DIFFERENT OF DIFFERENT SUBJECTS TO PERFORM DIFFERENT QUANTUM TASK, THROUGH USE OF ULTRAVIOLET'S ELECTRIC ENERGIES. ULTRAVIOLET CAN BE USED AS ELECTRIC ENERGY PROVIDER FOR INTERNAL ELECTRON NUCLEON NANO

LOCATIONS. WHO ARE IN NEEDS OF LESS POWERFUL ELERCTIC ENERGIES.

THE EARTH'S BIO FRIENDLY NON DISCOVERED FUNDAMENTAL PARTICLES

MOST OF THE EARTHS BIO FRIENDLY FUNDAMENTAL PARTICLES HAVE NOT BEEN DISCOVERED YET.

THE NON DISCOVERED BIO FRIENDLY FUNDAMENTAL PARTICLE CLASSES ARE BUILDING BLOCKS OF THE MOST OF EARTHS ELECTRONS, NUCLEONS AND ATOMS FUNDAMENTAL PARTICLE COMPOUND CONSTRUCTIONS COMPONENTS. THE NON DISCOVERED BIO FRIENDLY FUNDAMENTAL PARTICLES CLASSES (ND-BF-FP) OF EARTH ARE CREATORS OF THE EARTH ELEMENTS UNDER MOLECULAR EVOLUTION ORDERS AND PATHS. THE ND-BF-FP CONSTRUCT MOST OF LIVE AND NON LIVE ELECTRONS, PROTONS, AND NEUTRON'S FUNDAMENTAL PARTICLE MOLECULAR COMPOUND CONSTRUCTIONS, THE DISCOVERED FUNDAMENTAL PARTICLES ONLY ARE A

SEGMENTS FROM THESE PARTICLE MOLECULAR CONSTRUCTIONS.

EARTH'S BIO HOSTILE FUNDAMENTAL PARTICLES

THE MOST OF BIO HOSTILE FUNDAMENTAL PARTICLES AT EARTH ARE INTERNAL ELECTRON, NUCLEON FUNDAMENTAL PARTICLE COMPOUND CONSTRUCTION FORMS, THE FREE AIRBORNE BIO HOSTILE FUNDAMENTAL PARTICLE CLASSES AT ATMOSPHERE ARE RARE. THE LOCATIONS OF THE VICIOUS HIGH ENERGY POWERFUL THERMAL FUNDAMENTAL PARTICLES, ELECTRIC FUNDAMENTAL PARTICLES, LIGHT FUNDAMENTAL PARTICLES, AND SONIC VICIUOS FUNDAMENTAL PARTICLE CLASSES MOSTLY ARE IN CENTRAL PLANETARY LOCATION, THESE POWERFUL VICIOUS BIO HOSTILE FUNDAMENTAL PARTICLE CLASSES CONSTRUCT MOST OF THE INTERNAL ELECTRONS, INTER NUCLEONS PARTICLE MOLECULAR COMPOUND CONSTRUCTIONS OF EARTH'S CENTERAL LOCATION ATOMS.

SOME CLASSES FROM BIO HOSTILE FUNDAMENTAL PARTICLES, IN EARTH ARE SUCH AS:

1 - HEAVY WEIGHT FUNDAMENTAL PARTICLES (HEAVY MASS - FP), HIGH GRAVITON FORCE, HIGH THERMAL ENERGY CONTENT, POWERFULL BIO HOSTILE THERMAL FUNDAMENTAL PARTICLE CLASSES.

2 - HEAVY WEIGHT (HIGH MASS – FP), HIGH GRAVITON FORCE, HIGH ELECTRIC ENERGY CONTENT, BIO HOSTILE

VICIOUS POWERFUL, ELECTRIC FUNDAMENTAL PARTICLES FUNDAMENTAL PARTICLE CLASSES.

3 - X – FUNDAMENTAL PARTICLES CLASSES.

4 - GAMMA FUNDAMENTAL PARTICLE CLASSES.

5 - LASER AND OTHER COMPOSITE FUNDAMENTAL PARTICLE CLASSES,

6 - ETC.

THE GENERAL CHEMISTRY, PHYSICS OF ALL OF THESE BIO HOSTILE DIFFERENT FUNDAMENTAL PARTICLE CLASSES, ALL FOLLOW EXACTLY SAME CHEMISTRY AND PHYSICS LAWS AND ORDERS OF THE MATTER (FUNDAMENTAL PARTICLE) AS EXPLAINED IN THIS BOOK FOR DIFFERENT BIO FRIENDLY FUNDAMENTAL PARTICLES CLASSES. AS WELL ARE EXPLAINED IN OTHER AUTHOR'S BOOKS BRIEFLY.

THE FUNDAMENTAL PARTICLES GENERAL CHEMISTRY, ORGANIC CHEMISTRY, PHYSICS RULES AND LAWS FOR BOTH BIO HOSTILE AND BIO FRIENDLY PARTICLES ALL ARE EXACTLY SAME.

THE BIO HOSTILE FUNDAMENTAL PARTICLES AIRBORNE, OR TRAVEL IN P.A.S. PARTICLE CURRENT SYSTEMS TRANSIT IN SOLIDS, LIQUIDS OR PLASMA AND FROM THERE ENTER INTO INTERNAL ELECTRON, NUCLEON PCS OF ATOMS, THE BIO HOSTILE PARTICLES INSIDE ELECTRONS, NUCLEONS SUBSYSTEM UNITS CHEMICAL LAB.S, UNDER REGENERATIVE A. S. I. FP. Mol. CIC. COMBINE WITH RECIPIENT ELEMENTS INTERNAL ELECTRON NUCLEON PARTICLE COMPOUNDS, PRODUCE INTER ELECTRON NUCLEON BIO HOSTILE NEW

STRUTURE FUNDAMENTAL PARTICLE COMPOUNDS CONSTRUCTIONS.

THIS IS PHENOMENON OF BIO HOSTILE NEO ELECTRON GENESIS, NEO NUCLEON GENESIS UNDER THE A. S. I. FP. Mol. CIC. THIS PROCESS IS ALSO STORAGE OF BIO HOSTILE FUNDAMENTAL PARTICLES INSIDE ATOM, IN THE FORMS OF THE PARTICLE MOLECULAR COMPOUND CONSTRUCTIONS FORMS.

THE STORED BIO-CIDAL FUNDAMENTAL PARTICLES IN PARTICLE COMPOUND FORMS INSIDE ELECTRONS, AND NUCLEONS, CAN BE RELEASED AS FREE PARTICLES TO OUTSIDE ELECTRONS AND NUCLEONS, UNDER REVERSE - REGENERATIVE (OR DEGENERATIVE) A. S. I. FP. Mol. CIC. AGAIN.

THIS PHENOMENON IS RETRIEVAL PROCESS OF BIO HOSTILE FUNDAMENTAL PARTICLES TO OUT SIDE ELECTRONS, NUCLEONS AND ATOMS BY BREAKING DOWN THE LARGE INTERNAL ELECTRON, NUCLEON MOLECULAR FUNDAMENTAL PARTICLE COMPOUND CONSTRUCTIONS INTO SMALL SIZE PARTICLE MOLECULES UNDER RVERSE REGENERATIVE A. S. I. FP. Mol. CIC.

THE RELEASED BIO HOSTILE PARTICLE OF EARTH, PRODUCE POWERFUL INUSTRIAL THERMAL FUNDAMENTAL PARTICLE CURRENTS, ELECTRIC FUNDAMENTAL PARTICLE FLOWS, GAMMA PARTICLECURRNTS, AND X- FUNDAMENTAL PARTICLE CURRENTS, WHICH ALL OF THESE VICIOUS BIO – CIDAL FUNDAMENTAL PARTICLE CURRENTS PRESENTLY ARE USED IN DIFFERENT, COMMERCIAL, INDUSTRIAL,

THERAPEUTICAL, AND DIAGNOSTIC INDUSTRIES AND SCIENCES.

EARTH'S BIOFRIENDLY SENSIBLE SONIC FUNDAMENTAL PARTICLES (S – FP)

PLANET EARTH'S SONIC FUNDAMENTAL PARTICLES ARE LIGHT WEIGHT (LOWER QUANTUM MASS QUANTITY SONIC FUNDAMENTAL PARTICLES), LESS SONIC ENERGY CONTENT, AND LESS QUANTUM GRAVITON FORCE CONTENT, MOSTLY BIO FRIENDLY, BENIGN AND WEAK SOUND FUNDAMENTAL PARTICLE (S – FP) CLASSES.

IN EARTH, DIFFERENT SUBJECTS EMIT SUBJECT SPECIFIC SOUND FUNDAMENTAL PARTICLE CLOUD FROM SELF INTO THE AIR, OR TRANSMIT SONIC PARTICLES INTO P.A.S. MEDIA, WHICH THESE SONIC FUNDAMENTAL PARTICLE CLOUDS ALL ARE SUBJECT SPECIFIC PARTICLE COMPOSITES, PARTICLE INFORMATION IMAGE CLOUDS, OR FUNDAMENTAL PARTICLE COLONIES.

THE EMITTED SONIC FUNDAMENTAL PARTICLE INFORMATION, IMAGE PARTICLE CLOUDS (S- FP – I.I. – P. Cl.) TRAVEL AIRBORNE AND ENTER INTO EXTERNAL AUDITARY EAR CANALS OF LIVING THINGS, OR DIRECT

ENTRY INTO RCIPIENT ATOMS INTERAL ELECTRON NUCLEON PCS.

THE SENSARY AUDITARY ELECTRONS, NUCELONS OF INNER EAR, THROUGH POWERFUL ATTRACTIVE GRAVITON FORCES OF THEIR ATOMS, ATTRACT AIRBORNE EXOGENOUS SONIC PARTICLE CLOUDS, TRANSIT THE SOUND FUNDAMENTAL PARTICLE CLOUDS INTO CNS INTERNAL ELECTRON NUCLEONS PCS OF THE CNS AUDITARY CENTER. THE LIVING THINGS CNS AUDITARY CENTERS ELECTRONS AND NUCLEONS ARE IN NEEDS OF THESE PARTICLE CLOUDS TO USE THOSE SONIC FUNDAMENTAL PARTICLES IN ORDER TO CONSTRUCT THEIR INTERNAL ELECTRON NUCLEON PARTICLE COMPOUND MOLECULAR CONSTRUCTIONS, THROUGH REGNERATIVE A. S. I. FP. Mol. CIC.

TRANSMITTED EXOGENOUS SONIC FUNDAMENTAL PARTICLE INFORMATION AND IMAGE PARTICLE CLOUDS (S – FP - I.I. - P. Cl.) THROUGH INTERNAL ATOMS PCS CURRENTS, TRAVEL INTO CNS INTERNAL ELECTRON, NUCLEON SUBSYSTEM UNITS CHEMICAL LAB.S, UNDER REGENERATIVE A. S. I. FP. Mol. CIC. COMBINE WITH CNS INTERNAL ELECTRON NUCLEON PARTICLE COMPOUNDS, CONSTRUCT INTERNAL ELECTRON, NUCLEON SONIC PARTICLE COMPOUND NEO MOLECULAR CONSTRUCTIONS THROUGH USAGE OF INCOMING SONIC PARTICLE CLOUDS.

THIS IS AUDITARY CNS INTERNAL ELECTRON, NUCLEONS NEO PARTICLE COMPOUND GENESIS, THROUGH STORING EXOGENOUS S. FP. I.I. P. Cl. IN FORM OF SONIC FUNDAMENTAL PARTICLE INFORMATION CLOUD COMPOUNDS, INSIDE CNS ELECTRONS, NEUTRONS, AND

PROTONS. THIS IS PROCESSES OF STORING ENVIRONEMENTAL KNOWLEDGES ABOUT DIFFERENT SUBJECTS, AND SCIENCES, IN PARTICLE CLOUD COMPOUND FORMS, INSIDE CNS ATOM. (NEO PSYCHE GENESIS PHENOMENON OF ELECTRONS, NUCLEONS)

THESE SONIC PARTICLE CLOUD COMPOUNDS ARE EXACT SONIC PARTICLE COPIES OF DIFFERENT EXOGENOUS SUBJECTS, THAT RECORDED INSIDE CNS ELECTRONS AND NUCLEONS, AND ARE ENCYCLOPEDIA OF THE OUTSIDE INFORMATIONS AND IMAGES, STORED INSIDE CNS ATOMS.

THIS IS PHENOMENON OF SONIC PARTICLE COMPOUND NEO GENESIS THROUGH COMBINATION WITH EX. S. - FP – I.I. – P. Cl. ALSO CONSTRUCT NEW ELECTRONS AND NUCLEONS CONSTRUCTIONS. THIS PROCESS ALSO IS STORAGE OF EXOGENOUS SONIC PARTICLE CLOUDS INSIDE THE CNS ATOMS, WHICH IT STORE ENVIRONMENTAL OUSIDE INFORMATION CLOUDS, KNOWLEDGE AND SCIENCE INSIDE ATOMS, UNDER A. S. I. FP. Mol. CIC.

LIVING THINGS MEMORIZATION PHENOMENONS

RETRIEVAL OF EX. STORED FUNDAMENTAL PARTICLES INFORMATION IMAGE PARTICLE CLOUDS

THE EXOGENOUS SONIC FUNDAMENTAL PARTICLE INFORMATION AND IMAGE PARTICLE CLOUDS, LIGHT FUNDAMENTAL PARTICLE INFORMATION IMAGE PARTICLE CLOUDS (S. Y. – F.P. – I.I. – P. cl.), WITH EX. THERMAL - ELECTRIC FUNDAMENTAL PARTICLES INFORMATIONS AND IMAGE PARTICLE CLOUDS (T. E. – FP – I.I. – P. Cl.), UNDER A. S. I. FP. Mol. CIC. COMBINE WITH INTERNAL ELECTRON NUCLEON PARTICLE COMPOUND MOLECULES, PRODUCE INTERNAL ELECTRON NUCLEON S. Y. T. E. – FP – I.I. – P. Cl. – COMPOUNDS, CONSTRUCT PARTICLE CLOUD COMPOUND CONSTRUCTIONS OF CNS ELECTRONS, NUCLEONS. THIS IS PHENOMENON OF STORAGE OF EXOGENOUS ENVIRONMENTAL INFORMATIONS, IMAGES, AND KNOWLEDGE (S. Y. T.E. - FP- I.I. – P. Cl. COMPOUND FORMS) AT INSIDE CNS ELECTRONS, NUCLEONS.

BIOLOGICALLY THE INTERACTIONS OF THESE PARTICLE CLOUDS WITH EACH OTHER SENSE AS THOUGHT CURRENTS AND PSYCHE. THIS IS F.P. –I.I.- P.cl. STRORAGE PHENOMENON.

THE COMPLEX LARGE S. Y. T. E. – FP. – I.I. – P. Cl. MOLECULAR COMPOUNDS INSIDE INTERNAL ELECTRON, NUCLEON SUBSYSTEM UNIT CHEMICAL LAB.S, UNDER THE DEGENERETIVE A. S. I. FP. Mol. CIC. BREAK DOWN INTO SMALLER CONSTRUCTING FREE PARTICLE CLOUD MOLECULAR FORMS.

EACH GIVEN INDIVIDUALLY RELEASED S. Y. T. E. - FP – I.I – P. Cl. BY ITSELF IS REPRESENTATIVE OF SPECIFIC PAST ENVIRONMENTAL EVENTS, INFORMATION AND IMAGES. THAT FREED INFORMATION CLOUDS, AND IMAGE PARTICLE

CLOUDS AND THEIR INTERACTIONS UNDER A. S. I. FP. Mol. CIC. WITH EACH OTHER BIOLOGICALLY FEEL AND SENSES PAST SCENERIES AGAIN. THIS IS MEMORIZATION, AND RETRIEVAL PHENOMENONS.

NON SENSIBLE, BIO FRIENDLY, NON DISCOVERED FUNDAMENTAL PARTICLE CLASSES OF PLANET EARTH

(NS-BF-ND-FP)

NUMBER OF NON DISCOVERED FUNDAMENTAL PARTICLE CLASSES AT EARTH, ARE LARGER THAN NUMBERS OF ABOVE MENTIONED FEW BIO FRIENDLY FUNDAMENTAL PARTICLE CLASSES.

EARTH'S ATMOSPHERE DO NOT HAVE FREE HIGH ENERGY LETHAL, OR BIOHOSTILE FUNDAMENTAL PARTICLES. THIS FACTOR CAUSED SYNTHESIS OF LIVING THINGS INTERNAL ELECTRON, NUCLON PARTICLE STRUCTURES, THROUGH USING EXISTING WEAK ENERGY, BENIGN BIO FRIENDLY INDIGENOUS LIGHT PARTICLES, THERMAL AND ELECTRIC FUNDAMENTAL PARTICLES, ALSO USED WERE ALL OTHER KIND SONIC PARTICLES, OR MOSTLY CONSTRUCTED INTERNAL ATOM PARTICLE MOLECULAR CONSTRUCTIONS BY USING INDIGENOUS NS – BF – ND – FP FUNDAMENTAL PARTICLES AND CONSTRUCTED EARTH'S INTERNAL

ELECTRON, NUCLEON PARTICLE COMPOUND MOLECULAR CONSTRUCTIONS.

ULTRASONIC FUNDAMENTAL PARTICLE CLASSES ALSO EXIST IN EARTH. POWERFUL LASER MICROWAVE FUNDAMENTAL PARTICLES CLASSES OF EARTH, MOSTLY HAVE BEEN USED IN COMMERCIAL, MEDICAL OR NON MEDICAL INDUSTRIAL AND TELECOMMUNICATION, OR LOGESTICS INDUSTRIES. ALSO THERE ARE LARGE OTHER UNKNOWN SONIC FUNDAMENTAL PARTICLE CLASSES IN EARTH, WHICH ARE NON SENSIBLE TO MAN KIND, BUT SOME OF THESE PARTICLES ARE SENSIBLE TO OTHER LIVING SPECIES.

THE NON SENSIBLE, BIO FRIENDLY, NON DISCOVERED FUNDAMENTAL PARTICLES (NS- BF – ND – FP), EITHER AIRBORNE, OR THROUGH OTHER EXISTING PCS TRAVEL AND ENTER INTO INTERNAL ELECTRON, NUCLEON'S SUBSYSTEM UNITS CHEMICAL LAB.S, LIKE OTHER PARTICLES, UNDER A. S. I. FP. Mol. CIC. CONBINE WITH INTERNAL ELECTRON AND NUCLEON PARTICLE COMPOUND CONSTRUCTIONS OF LIVE OR NON LIVE ATOMS. AND BUILD INTERNAL ELECTRON, INTERNAL NUCLEON PARTICLE COMPOUND MOLECULAR CONSTRUCTIONS, SIDE BY SIDE SIMILAR TO BIO FRIENDLY FUNDAMENTAL PARTICLES OF EARTH PLANET.

THE LAWS AND RULES OF FUNDAMENTAL PARTICLES ALL OVER UNIVERSE ARE THE SAME, THEY ALL FOLLOW EXACT SAME CHEMICAL, PHYSICAL, BIOLOGICAL RULES AND LAWS AS DESCRIBED BRIEFLY IN THIS BOOK.

Quantum Energy content of Particles

Quantum Energy Donors and Quantum Energy Recipients

(PERFORMING PHYSICAL, CHEMICAL BIOLOGICAL TASK, INSIDE ELECTRONS, NUCLEONS BY USING ENERGY)

HERE INTERRELATIONSHIPS OF THERMAL ENERGY WITH INFRARED THERMAL PARTICLES BRIEFLY EXPLAINED, THE EXACT SAME INTERRELATIONSHIPS ALSO EXIST BETWEEN THE OTHER DIFFERENT KIND ENERGIES SUCH AS ELECTRIC ENERGY, SONIC ENERGY, LIGHT ENERGY, GAMMA ENERGY, ETC. WITH THEIR RELATED DIFFERENT OTHER KIND FUNDAMENTAL PARTICLES (SUCH AS ELECTRIC PARTICLES, SONIC PARTICLES, LIGHT PARTICLES, GAMMA PARTICLES, ETC.).

THERE ARE ALSO ELECTRIC ENERGY, SONIC ENERGY, X-ENERGY, GAMMA ENERGY, ETC. PROVIDER ATOMS, ELECTRONS, NEUTRONS, AND PROTONS. AS WELL AS THERE

ARE LIGHT ENEGRY, ELECTRIC ENERGY, SONIC ENERGY, THERMAL ENERGY, ETC. RECIPIENT ATOMS, ELECTRONS, NUCLEONS.

THE EXPLAINED INFRARED PARTICLES INTERRELATIONSHIPS RULES AND LAWS WITH THERMAL ENERGY RULES AND LAWS, APPLY EXACTLY THE SAME FOR ALL OTHER KINDS NANO STRUCTURES, SUCH AS NEUTRONS, ATOMS, PROTONS, ELECTRONS, ETC. AS WELL, WITH NO EXCEPTION.

INTERRELATIONSHIPS OF THERMAL ENERGY WITH INFRARED AND RED PARTICLES

INTERRELATIONSHIPS OF THERMAL ATOMS, OR THERMAL ELECTRON NUCLEONS WITH THERMAL ENERGIES

RED PARTICLE COMPOUNDS, AND INFRARED THERMAL FUNDAMENTAL PARTICLES STORE THERMAL ENERGY IN

FORMS OF PARTICLE COMPOUNDS INSIDE PROTONS, ELECTRONS, NEUTRONS. THIS IS PHENOMENON OF STORING THERMAL ENERGY IN PARTICLE COMPOUND FORMS INSIDE ATOMS.

THERMAL FUNDAMENTAL PARTICLES PROVIDE THERMAL ENERGY INTO DIFFERENT INTERNAL ELECTRON, NUCLEON NANO LOCATIONS, IN ORDER THROUGH USE OF THESE ENERGIES ACHIEVE AND CONTROL PHYSICAL BIOLOGICAL AND CHEMICAL NANO TASKS IN DIFFERENT NANO LOCATIONS.

THE ELECTRONS AND NUCLEON'S PIS THROUGH ISSUING PROPER PARTICLE CLOUD ORDERS REGULATE INTERNAL ELECTRONS, NUCLEONS QUANTUM TASK OPERATIONS. FOR EXAMPLE, IN CASES OF NEEDED A. S. I. FP. Mol. CIC. OPERATIONS, THE ELECTRONS, NUCLEON'S PIS EMIT ORDER – CLOUDS, WHICH THESE ORDERS TRANSMIT THROUGH INTERNAL ELECTRON NUCLEON PCS INTO DIFFERENT RELATED SUBSYSTEM UNIT'S CHEMICAL LAB.S.

UNDER THE ISSUED ORDER -CLOUDS INSIDE SUBSYSTEM UNITS CHEMICAL LAB.S, THE A. S. I. FP. Mol. CIC. EITHER STARTS, OR STOP, AND CONTROL THE INTERACTIONS ACCORDING PROVIDING NEEDED THERMAL ENERGIES, CONTINUING THE CORRECT TEMPERATURES TO ACHIEVE THE PARTICLE MOLECULAR COMBINATIONS, OR STOP NEEDED THERMAL ENERGIES IN ORDER TO STOP THE FUNCTION OF THE CHEMICAL INTERACTIONS, INSIDE SUBSYSTEM UNIT'S CHEMICAL LAB.S.

THE ATOM'S PIS CAN CONTROL, ENTIRE AND ALL ONGOING INTERNAL ELECTRON, NUCLEON PHYSICAL, MECHANICAL,

CHEMICAL, BIOLOGICAL QUANTUM FUNCTIONS, IN ANY GIVEN QUANTUM LOCATIONS. THROUGH ISSUED PARTICLE CLOUD ORDERS, EITHER STOP ONGOING QUANTUM TASKS OF ANY KIND, OR CONTINUE ACCORDING THE CLOUD-ORDERS.

PIS ISSUED ORDER CLOUDS CIRCULATE BETWEEN DIFFERENT ELECTRONS, NUCLEONS INTERNAL ATOM'S PARTICLE CIRCULATION SYSTEMS (PCS), THE PARTICLE CURRENTS OT THE INTERNAL ATOM PCS TRANSMIT AND CARRY THE PARTICLE CLOUD ORDERS INTO SELECTED NANO UNIT'S QUANTUM LOCATIONS. WHICH THE NEEDED QUANTUM FUNCTIONS OF MECHANICAL, PHYSICAL, BIOLOGICAL, OR CHEMICAL NANO TASKS MUST TAKE PLACE ACCORDING THE ISSUED CLOUD ORDERS. AT NEEDED QUANTUM LOCATIONS (NANO SITES).

THERMAL ENERGY INTERRELATIONSHIP RULES WITH INFRARED PARTICLES APPLY TO ALL OTHER DIFFERENT FUNDAMENTAL PARTICLE KINDS EXACTLY THE SAME

The Electric Energy, Thermal energy, Gamma energy, Light energy, Sonic, (etc.) Energy interrelationships Rules and Laws with (Electric -Thermal – Gamma – Light- Sonic) (Electrons – Neutrons – Protons and Atoms) all follow the same laws and orders similar to infrared and red particles. that explained for Infrared and Red particles in above.

THE ELECTRIC FUNDAMENTAL PARTICLES PROVIDE ELECTRIC ENERGY, IN QUANTUM QUANTITIES INTO DIFFERENT INTERNAL ELECTRON, INTERNAL NUCLEON NANO LOCATIONS, TO BE USED INSIDE DIFFERENT QUANTUM LOCATIONS FOR ACHIEVING DIFFERENT KIND PHYSICAL, CHEMICAL, MECHANICAL AND BIOLOGICAL NANO TASKS, INSIDE DIFFERENT NANO UNITS, THROUGH USE OF PROVIDED ELECTRIC ENERGIES.

LIGHT FUNDAMENTAL PARTICLES, SONIC FUNDAMENTAL PARTICLES AND ALL OTHER BIO FRIENDLY FUNDAMENTAL PARTICLES, AS WELL AS ALL BIO HOSTILE FUNDAMENTAL PARTICLES, ALSO THROUGH PROVIDING THEIR DIFFERENT KINDS QUANTUM ENERGIES TO DIFFERENT NANO SITE, ACHIEVE DIFFERENT PHYSICAL, CHEMICAL, BIOLOGICAL TASKS, INSIDE THE DIFFERENT ELECTRONS, PROTONS, NEUTRONS.

ALL OF DIFFERENT ELECTRONS, NEUTRONS, PROTONS, ALL NEEDS DIFFERENT KINDS ELECTRIC, THERMAL, SONIC, LIGHT, OR OTHER TYPES NANO ENERGIES, IN ORDER TO DO NANO TASKS, IN DIFFERENT QUANTUM LOCATIONS. (THE QUANTUM ENERGY RECIPIENTS, AND QUANTUM ENERGY DONOR PHENOMENONS).

QUANTUM UNIT THERMAL ENERGY
(ONE UNIT THERMAL ENERGY)

THE THERMAL ENERGY CONTENT OF ONE INFRARED FUNDAMENTAL PARTICLE IS EQUAL TO ONE QUANTUM THERMAL ENERGY UNIT. THE QUANTUM THERMAL ENERGY UNIT OR ANY OTHER QUANTUM LEVELS THE OTHER KIND ENERGY UNITS ARE USED IN MEASUREMENTS OF THE QUANTUM LEVEL CALCULATIONS.

THE EXISTING THERMAL ENERGY AMOUNT IN ONE INFRARED THERMAL FUNDAMENTAL PARTICLE IS EQUAL TO ONE QUANTUM UNIT THERMAL ENERGY.

THERMAL INFRARED FUNDAMENTAL PARTICLE (T. Ir. – FP) IS A BIOFRIENDLY WEAK THERMAL ENERGY CONTENT FUNDAMENTAL PARTICLE IN EARTH, THE INFRARED SIMILAR TO OTHER THERMAL PARTICLES PROVIDES QUANTUM THERMAL ENERGY INTO INTERNAL ELECTRON NUCLEON QUANTUM LOCATIONS (NANO SITES), TO BE USED TO ACHIEVE DIFFERENT BIOLOGICAL CHEMICAL AND PHYSICAL NANO TASKS, THROUGH THE USE OF PARTICLE PROVIDED THERMAL ENERGIES.

THERMAL INFRARED FUNDAMENTAL PARTICLE COMPOUNDS (T. Ir. – FP - COMP.) SIMILAR TO OTHER PARTICLES HAVE BEEN USED IN CONSTRUCTIONS OF DIFFERENT INTERNAL ELECTRON, NUCLEON PARTICLE COMPOUND STRUCTURES (THIS IS PHENOMENON OF THERMAL ENERGY STORAGE PHENOMENON INSIDE PARTICLE COMPOUNDS).

UNDER DEGENERATIVE A. S. I. F.P. Mol. C.I.C. THESE LARGER MOLECULAR THERMAL FUNDAMENTAL PARTICLE COMPOUNDS CAN BREAK DOWN INTO SMALLER MOLECULAR PARTICLE MOLECULES, AND RELEASE INFRARED FUNDAMENTAL PARTICLES FREE, INTO INTERNAL ELECTRON, NUCLEON NANO SITES, AS THERMAL ENERGY PROVIDER IN ANY GIVEN HEAT ENERGY NEEDED NANO LOCATIONS.

THERMAL ELECTRONS, THERMAL NUCLEONS

DIFFERENCES OF RED PARTICLE'S A. S. I. FP. Mol. CIC. WITH INFRARED A. S. I. FP. Mol. CIC. INSIDE ATOMS

THE CHEMICAL INTERACTION DIFFERENCES OF INFRARED PARTICLES FROM RED PARTICLES

The Red Particles as well as Infrared Particles, both travel airborne, enter into Electrons, Nucleons internal Subsystem Unit's Chemical Lab. through internal Atom's PCS. These exogenous fundamental particles under Regenerative A. S. I. FP. Mol. CIC. combine with internal Electron, Nucleon particle compounds, construct different molecular infrared or red Particle Compound Construction Electron and Nucleons. (PHENOMENON OF NEO PARTICLE COMPOUND GENESIS WITH RED AND INFRARED PARTICLES).

RED COLOR LIGHT FUNDAMENTAL PARTICLE (Y. r. –F.P.) OR RED COLOR THERMAL FUNDAMENTAL PARTICLE (T. r. –

F.P.), AND INFRARED THERMAL FUNDAMENTAL PARTICLES (T. Ir. – FP), POSSESSES TWO DIFFERENT KIND BIOFRIENDLY THERMAL ENERGY CONTENTS, AND BIOFRIENDLY LIGHT ENERGY CONTENTS.

THE RED FUNDAMENTAL PARTICLE'S LIGHT ENERGY CONTENT IS MORE THAN THE INFRARED FUNDAMENTAL PARTICLE'S LIGHT ENERGY QUANTITY. AT THE SAME TIME, THE THERMAL ENERGY CONTENTS OF THE INFRARED FUNDAMENTAL PARTICLES, EXCEED THE THERMAL ENERGY QUANTITY OF THE RED COLOR FUNDAMENTAL PARTICLES.

THE RED FUNDAMENTAL PARTICLE'S WEIGHT (QUANTUM Yr. - FP MASS QUANTITY IN ONE FUNDAMENTAL PARTICLE), IS DIFFERENT THAN THE PARTICLE'S WEIGHT OF THE INFRARED FUNDAMENTAL PARTICLE'S QUANTUM MASS QUANTITIES IN ONE FUNDAMENTAL PARTICLE.

THE INFRARED FUNDAMENTAL PARTICLE'S GRAVITON FORCES, ARE DIFFERENT THAN THE GRAVITON FORCES OF RED COLOR FUNDAMENTAL PARTICLES, THEIR FUNDAMENTAL PARTICLE SHAPES, COLORS, WAVE LENGTH, PHYSICAL, CHEMICAL, BIOLOGICAL PROPERTIES OF THESE TWO FUNDAMENTAL PARTICLES ARE DIFFERENT FROM EACH OTHER.

THEREFORE, THE CONSTRUCTED RED FUNDAMENTAL PARTICLE COMPOUNDS INSIDE ELECTRONS AND NUCLEONS ALSO ARE DIFFERENT THAN FROM INFRARED PRODUCED INFRARED PARTICLE MOLECULAR COMPOUND CONSTRUCTIONS.

THE END RESULTS ARE PRODUCTIONS OF THE DIFFERENT KIND NEO PARTICLE COMPOUND MOLECULAR CONSTRUCTIONS, DIFFERENT KIND NEO ELECTRON GENESIS, DIFFERENT KINDS OF NEO NUCLEON GENESIS AT END RESULTS. RED FUNDAMENTAL PARTICLES REGENERATIVE AND DEGENERATIVE A. S. I. FP. Mol. CIC. ALSO ARE DIFFERENT FROM INFRARED FUNDAMENTAL PARTICLE'S A. S. I. FP. Mol. CIC.

THERMAL ATOMS MOSTLY CONSTRUCTED FROM DIFFERENT KIND THERMAL FUNDAMENTAL PARTICLE MOLECULAR CONSTRUCTIONS. AND ELECTRIC FUNDAMENTAL PARTICLE CONSTRUCTION ATOMS MOSTLY ARE MADE FROM DIFFERENT KIND ELECTRIC PARTICLE MOLECULAR COMPOUND CONSTRUCTIONS. ETC. SO ON.

A. S. I. FP. Mol. CIC
ESSENTIAL RULES AND LAWS OF FUNDAMENTAL PARTICLES CHEMISTRY

AUTONOMOUS SEQUENTIAL INTER FUNDAMENTAL PARTICLE CHEMICAL INTERACTION CYCLES & CHAINS

ONE NORMAL PARTICLE MOLECULE (NPM)

ONE NPM IS A GIVEN WELL DEFINED, FIXED, CERTAIN, CONSTANT, KNOWN CHEMICAL FORMULARY FUNDAMENTAL PARTICLE MOLECULE, WHICH IT'S NPM MASS, ENERGY KIND AND ENERGY QUANTITY, AND GRAVITON FORCE QUANTITIES ALL ARE SPECIFIC FOR THAT GIVEN MOLECULE PARTICLE ONLY. DIFFERENT FUNDAMENTAL PARTICLES POSSESSES DIFFERENT NORMAL PARTICLE MOLECULES PROPERTIES.

IN ALL A. S. I. FP. Mol. CIC. BETWEEN TWO COMBINING FUNDAMENTAL PARTICLES, ALWAYS ONE NORMAL FUNDAMENTAL PARTICLE MOLECULE (NPM) FROM ONE FUNDAMENTAL PARTICLE, COMBINE WITH ONE ANOTHER COMPATIBLE ONE NPM FROM ANOTHER KIND FUNDAMENTAL PARTICLE MOLECULE, UNDER REGENERATIVE OR DEGENERATIVE A. S. I. FP. Mol. CIC. AND PRODUCE ONE ANOTHER KIND NPM GIVEN WELL DEFINED FIX CHEMICAL FORMULARY TERTIARY ONE NORMAL FUNDAMENTAL PARTICLE COMPOUND.

IN ALL A. S. I. FP. Mol. CIC. BETWEEN COMBINING COMPATIBLE TWO ATTRACTIVE GRAVITON FORCE FUNDAMENTAL PARTICLE MOLECULES IN ANY GIVEN CHEMICAL INTERACTION TESTS, ALWAYS ONE DOMINANT GRAVITON FORCE FUNDAMENTAL PARTICLE MOLECULE THROUGH IT'S DOMINANT HIGH ATTRACTIVE GRAVITON FORCE TAKES NUCLEAR POSITION, AND ATTRACT THE OTHER COMPATIBLE LESSER GRAVITON FORCE RECESSIVE ONE NPM FUNDAMENTAL PARTICLE FROM PERIPHERY, THE TWO ATTRACTING DOMINANT AND RECESSIVE FUNDAMENTAL PARTICLES COMBINE TO EACH OTHER, AND

PRODUCE ONE GIVEN WELL DEFINED NPM OTHER DIPLO FUNDAMENTAL PARTICLE MOLECULE COMPOUND.

IN THESE CHEMICAL COMBINATIONS ALWAYS DOMINANT FUNDAMENTAL PARTICLES OCCUPY NUCLEUS LOCATION, AND THE RECESSIVE FUNDAMENTAL PARTICLES STAY AT PERIPHERAL COMPARTMENT.

WHEN COMBINATION OF ALL EXISTING TWO GROUPS ATTRACTING FUNDAMENTAL PARTICLE POPULATION IN ANY KIND GIVEN CHEMICAL TEST TOTALLY ONE BY ONE COMBINED WITH EACH OTHER ENTIRELY. AT THIS STAGE THE TERMINATION POINT OF INTERACTION ACCOMPLISHED AND A. S. I. FP. Mol. CIC. IS DONE AND COMPLETED.

UNDER REPULSIVE GRAVITON FORCES OF THE FUNDAMENTAL PARTICLE, THE MOLECULAR PARTICLES REPULSE EACH OTHER, AND REFUSE COMBINATION, AND A. S. I. FP. Mol. CIC. REFUSE TO TAKE PLACE.

IN A. S. I. FP. Mol. CIC. ALWAYS MULTIPLE CHAINS OF INDIVIDUAL DIFFERENT SINGLE CHAIN CHEMICAL INTERACTIONS ONE AFTER OTHER TAKE PLACE SEQUENTIALLY, AUTONOMOUSLY WITH EXPONENTIAL CHEMICAL INTERACTION SPEEDS, IN CYCLES OR CHAINS. AND PRODUCE NUMEROUS DIFFERENT MOLECULAR SEQUENTIAL MOLECULAR PARTICLE COMPOUND CONSTRUCTIONS, AUTONOMOUSLY UNDER THE PAST WORLD HISTORY EXACT MOLECULAR EVOLUTION PATHS AND ORDERS COPIES. UNTIL THE COMPLETION ONE REGENERATIVE OR DEGENERATIVE A. S. I. FP. Mol. CIC. IS ENTIRELY ACHIEVED.

IN FINAL CONSEQUENCES OF THESE A. S. I. FP. Mol. CIC. ALWAYS AT ALL WELL DEFINED, CERTAIN AND CONSTANT MOLECULAR COMPOUNDS NANO AND MICRO CONSTRUCTIONS TAKE PLACE AND COMPLETE IT'S CONSTRUCTION ORDERS AND GOALS, ACCORDING THE MOLECULAR EVOLUTION PATHS OF THE PAST WORLD MOLECULAR EVOLUTION HISTORY ORDERS.

INTRODUCTION TO FUNDAMENTAL PARTICLE BIOCHEMISTRY

Chemical Combination Between two different Fundamental Particle Molecules in Flower under A. S. I. FP. Mol. CIC

IN PLANTS REGARDING THE A. S. I. FP. Mol. CIC. BETWEEN THE AIRBORNE, SUN ORIGIN, EXOGENOUS FUNDAMENTAL PARTICLES (EX. FP) CHEMICAL COMBINATIONS, WHEN COMBINING WITH INTERNAL PLANT'S ELECTRONS, NUCLEONS PARTICLE COMPOUNDS MOLECULES, THE EXAMPLE HERE IS COMBINATIONS OF ONE NPM RED LIGHT FUNDAMENTAL PARTICLE WITH ONE NPM FLOWER

INTERNAL ELECTRON NUCLEON PARTICLE COMPOUND MOLECULAR STRUCTURE, WHICH PRODUCE ONE NPM RED LIGHT PARTICLE COMPOUND STRUCTURE INSIDE FLOWER ATOMS.

RED COLOR LIGHT FUNDAMENTAL PARTICLES TRAVEL AIRBORNE FROM PLANET SUN, DIRECTLY FROM FLOWERS PAS ENTER INTO INTERNAL ATOM PCS, THROUGH INTERNAL ELECTRON – NUCLEON PCS TRANSIT INTO ATOM'S SUBSYSTEM UNITS CHEMICAL LAB.S, AND COMBINE WITH INTERNAL ELECTRON, NUCLEON PARTICLE COMPOUNDS, PRODUCE RED LIGHT PARTICLE COMPOUND UNDER REGENERATIVE A. S. I. FP. Mol. CIC. THE CHEMICAL COMBINATION OF ONE NORMAL MOLECULE WEIGHT GIVEN EXOGENOUS AIRBORNE RED LIGHT FUNDAMENTAL PARTICLE MOLECULE COMBINE WITH ONE NPM FLOWER'S INTERNAL ELECTRON, NUCLEON PARTICLE COMPOUNDS. PRODUCE TERTIARY ONE NPM RED LIGHT FUNDAMENTAL PARTICLE COMPOUND MOLECULES INSIDE FLOWER ATOMS. RED COLOR PARTICLE COMPOUNDS CONSTRUCT RED COLOR ELECTRONS, NUCLEONS, ATOMS, AND RED COLOR CELLS, FLOWERS.

THESE CHEMICAL INTERACTION OF ONE NPM RED LIGHT PARTICLE WITH ONE NPM INTERNAL ELECTRON, NUCLEON PARTICLE COMPOUND MOLECULE, ALWAYS PRODUCE ONE NORMAL MOLECULE RED PARTICLE COMPOUND INSIDE ELECTRONS AND NUCLEONS, (STORAGE OF RED LIGHT PARTICLE, WHICH HAS THERMAL ENERGY AND LIGHT ENERGY CONTENTS BOTH, NEEDED FOR INTERNAL ATOM AND INTERNAL CELLULAR PHYSICAL, CHEMICAL,

BIOLOGICAL QUANTUM TASKS, IN DIFFERENT QUANTUM LOCATIONS.) (PHENOMENON OF STORAGE OF PARTICLES).

REVERSE OF ABOVE REGENERATIVE A. S. I. FP. Mol. CIC. CAUSE RETRIEVAL OF THERMAL PARTICLES FREE IN INTERNAL ATOM OR OUTSIDE ATOM SPACE, AND PROVIDE NEEDED PARTICLES AND THEIR EXISTING ENERGY KIND TO DIFFERENT NANO – SITES, TO ACHIEVE DIFFERENT QUANTUM NANO TASKS, THROUGH USE OF THESE FUNDAMENTAL PARTICLES AND THEIR RELEASED ENERGIES. THIS IS PHENOMENON OF RETRIEVAL OF QUANTUM ENERGIES, AND RETRIEVAL OF FUNDAMENTAL PARTICLE CLOUDS.

REGARDING INFRARED A. S. I. FP. Mol. CIC., ALWAYS ONE EXOGENOUS NORMAL MOLECULE WEIGHT INFRARED PARTICLE, ENTER INTO INTERNAL ELECTRON, NUCLEON CHEMICAL LABS, AND COMBINE WITH COMPATIBLE ONE NORMAL MOLECULE WEIGHT OTHER PARTICLE COMPOUNDS, AND PRODUCE ANOTHER TYPE CHEMICAL FORMULARY STRUCTURE NORMAL WEIGHT PARTICLE COMPOUNDS.

ABOVE ARE TWO DIFFERENT KIND A. S. I. FP. Mol. CIC. AND ARE ENTIRELY DIFFERENT FROM EACH OTHER.

THE INTERNAL ELECTRON, NUCLEON RED LIGHT PARTICLE COMPOUNDS ARE QUANTUM THERMAL ENERGY STORAGE SYSTEMS, THESE NANO STRUCTURES ARE QUANTUM

THERMAL AND LIGHT ENERGY PROVIDERS, DUE TO THEIR INHERITED ENERGY CONTENTS.

THERE ARE LARGE NUMBER OF NON DISCOVERED, NON SENSIBLE OTHER BIO FRIENDLY THERMAL FUNDAMENTAL PARTICLES, WHICH CONSTRUCT INTERNAL ELECTRON, AND INTERNAL NUCLEON THERMAL PARTICLE COMPOUND CONSTRUCTIONS, AND OPERATE DIFFERENT LIVING THINGS BIOLOGICAL, PHYSICAL, CHEMICAL QUANTUN TASKS, FOR DIFFERENT LIVING THINGS ORGANS, AND SYSTEMS, AND ARE VITAL FOR LIFE. BUT NOT DISCOVERED YET.

THERE ARE REMARKABLE CLOSE COOPERATIONS AND COORDINATIONS BETWEEN QUANTUM THERMAL ENERGY DONOR FUNDAMENTAL PARTICLE SYSTEMS, WITH BIO MOLECULAR CELL OPERATED THERMAL ENERGY DONOR SYSTEMS AND ORGANS. THESE SYSTEMS AND ORGANS EITHER FUNDAMENTAL PARTICLE BASE, OR BIOMOLECULAR BASE THERMAL PROVIDER ORGANS, ARE IN CHARGE OF LIVING THINGS PHYSICAL, CHEMICAL, BIOLOGICAL FUNCTION OPERATIONS.

THERE ARE ALSO LARGE NUMBER OF DIFFERENT SENSIBLE THERMAL ENERGY, ELECTRIC ENERGY, LIGHT ENERGY PROVIDER OTHER COLOR AND CLASS FUNDAMENTAL PARTICLES, WHICH THEIR BRILLIANT, BRIGHT COLORS ARE OUT STANDING, AND PARTICIPATE IN CONSTRUCTION OF THE ELECTRONS AND NUCLEON, STILL HAVE NOT BEEN DISCOVERED YET PRESENTLY.

Chemical Combination between Blue Light Particle and Plant's internal Atom Particle Compounds under:

A. S. I. FP. Mol. CIC IN PLANTS

THE BLUE COLOR LIGHT – ELECTRIC FUNDAMENTAL PARTICLE, VIOLET COLOR ELECTRIC AND LIGHT FUNDAMENTAL PARTICLES, AS WELL AS MANY OTHER CLASS NON DISCOVERED BIO FRIENDLY AIRBORNE SENSIBLE OR NON SENSIBLE ELECTRIC AND LIGHT FUNDAMENTAL PARTICLES, DAILY TRAVEL AIRBORNE FROM PLANET SUN, DIRECTLTY ENTER INTO FLOWERS, PLANT'S INTERNAL ELECTRON, NUCLEON PCS. THESE EXOGENOUS BLUE, VIOLET, OR ANY OTHER CLASS LIGHT – ELECTRIC FUNDAMENTAL PARTICLES INSIDE PARTICLE SUBSYSTEM UNIT CHEMICAL LAB.S OF PLANTS ELECTRONS AND NUCLEONS COMBINE WITH INTERNAL ELECTRON, NUCLEON PARTICLE MOLECULAR COMPOUNDS, PRODUCE BLUE FUNDAMENTAL PARTICLE COMPOUNDS, VIOLET PARTICLE COMPOUNDS, ETC. UNDER REGENERATIVE A. S. I. FP. Mol. CIC. AND TINT THE FLOWER'S ELECTRONS, NUCLEONS, ATOMS AND CELL AS BLUE OR VIOLET, OR ANY OTHER DIFFERENT COLORS.

BIO FRIENDLY ELECTRIC ENERGY CONTENT, LIGHT ENERGY CONTENT, THERMAL ENERGY CONTENT, OR ANY OTHER KIND QUANTUM ENERGY QUANTITY CONTENT FUNDAMENTAL PARTICLES FOLLOWING A. S. I. FP. Mol. CIC. STORE THEIR EXISTING INHERITANCE DIFFERENT ENERGY KINDS, WITH DIFFERENT QUANTUM QUANTITY ENERGY CONTENTS INSIDE THE PLANTS INTERNAL ELECTRON, NUCLEONS IN FUNDAMENTAL PARTICLE COMPOUND FORMS.

THIS IS PHENOMENON OF STORAGE OF THE QUANTUM QUANTITY DIFFERENT ENERGY CONTENTS, INSIDE ATOM, IN PARTICLE MOLECULAR COMPOUND FORMS.

PROVIDING QUANTUM QUANTITY ELECTRIC ENERGY, SONIC ENERGY, THERMAL ENERGY, OR LIGHT ENERGIES FOR LIVING THINGS INTERNAL ELECTRON-NUCLEON QUANTUM LOCATIONS, FOR ACHIEVEMENTS OF DIFFERENT KIND NANO TASKS IN BIOLOGICAL, PHYSICAL, CHEMICAL OPERATIONS, ALL ARE FUNCTIONS OF THESE STORED FUNDAMENTAL PARTICLES.

THE BLUE COLOR LIGHT – ELECTRIC ENERGY CONTENT PARTICLE COMPOUNDS, THE VIOLET COLOR ELECTRIC – LIGHT ENERGY CONTENT FUNDAMENTAL PARTICLE COMPOUNDS, UNDER DEGENERATIVE A. S. I. FP. Mol. CIC. BREAK DOWN INTO SMALLER PARTICLE MOLECULAR STRUCTURES, AND PROVIDE LIGHT ENERGIES, ELECTRIC QUANTUM QUANTITY ENERGIES, AS WELL AS FREE BLUE OR VIOLET LIGHT PARTICLES AND ELECTRIC PARTICLES TO DIFFERENT NANO LOCATIONS, DELIVER NECESSARY NEEDED QUANTUM ELECTRIC AND LIGHT NERGIES, AND ACHIEVE

QUANTUM TASKS, IN DIFFERENT INTERNAL ELECTRONS, NUCLEON QUANTUM LOCATIONS. AND OPERATE BIOLOGICAL FUNCTIONS INSIDE DIFFERENT PLANTS INTERNAL ELECTRON NUCLEON PHYSICAL, CHEMICAL, MECHANICAL OPERATIONS.

THERE ARE ALSO LARGE NUMBER OF OTHER COLOR SNESIBLE NON DISCOVERED FUNDAMENTAL PARTICLES, WHICH THEIR MIXTURE CHEMICAL COMBINATIONS WITH ASSOCIATION OF BLUE LIGHT PARTICLES, OR OTHER COLOR LIGHT PARTICLES MIXTURES, CAUSE REMARKABLE INTERMEDIATE COLOR CHANGES IN FLOWERS AND PLANTS SPECIES. EVEN WHEN CONSIDERING ONE GIVEN COLOR LIGHT PARTICLE CHEMICAL COMBINATION, UNDER A. S. I. FP. Mol. CIC. ADDITIONALLY, GOING TO PRODUCE DIFFERENT INTERMEDIATE FLOWER COLORS AS WELL.

BOTH BLUE PARTICLE AND VIOLET PARTICLE POSSESSES LIGHT ENERGIES AND ELECTRIC ENERGIES IN DIFFERENT QUANTITIES IN FUNDAMENTAL PARTICLE CONTENTS, THE PARTICLE WEIGHTS AND THE PARTICLE GRAVITON FORCES OF THE BLUE COLOR LIGHT PARTICLE ARE DIFFERENT THAN THE VIOLET COLOR PARTICLE.

THE BLUE COLOR PARTICLE A. S. I. FP. Mol. CIC. IS DIFFERENT THAN THE VIOLET PARTICLES A. S. I. FP. Mol. CIC. ALSO THERE ARE MANY OTHER NON DISCOVERED, BIO FRIENDLY SENSIBLE AND NON SENSIBLE INTERMEDIATE DIFFERENT LIGHT FUNDAMENTAL PARTICLE CLASSES, WHICH THEIR EXISTENCES ARE NOT DISCOVERED AT PRESENT TIME. ALL OF THESE FUNDAMENTAL PARTICLES ALSO HAVE VITAL IMPORTANCES FOR PLANTS AND LIVING

THINGS EXISTENCES. LIFE WITHOUT THESE PARTICLES CAN NOT EXIST IN EARTH.

THERMAL ATOMS, ELECTRIC ATOMS

THERMAL ENERGY ATOMS AND ELECTRIC ENERGY ATOMS

(INTERNAL ELECTRON, NUCLEON ENERGY STORAGE SYSTEMS)

SONIC, THERMAL, LIGHT, ELECTRIC PARTICLE CONSTRUCTION ELECTRONS, NEUTRONS, PROTONS

Thermal – Light – Electric - Atoms, and Thermal – Electric – Light Atom constructed Bio Molecular Compounds

BIOMOLECULAR ENERGY CONSTRUCTIONS (INTER - CELL ENERGY STORAGE SYSTEMS)

FUNDAMENTAL PARTICLES POSSESS PARTICLE SPECIFIC ENERGY KIND AND ENERGY CONTENT. EACH GIVEN PARTICLE HAVE WELL DEFINED, CERTAIN KINDS, RELATIVELY FIX QUANTITIES QUANTUM ENERGY CONTENT. THE ENERGY CONTENTS OF DIFFERENT FUNDAMENTAL PARTICLES AT DIFFERENT CLASSES ARE DIFFERENT FROM EACH OTHER.

EACH GIVEN ELECTRON, NEUTRON, PROTON, ATOM HAVE BEEN CONSTRUCTED FROM WELL DEFINED DIFFERENT KIND FUNDAMENTAL PARTICLE MOLECULAR COMPOUNDS. THE ENERGY QUANTITY CONTENTS OF THESE ELECTRONS, NUCLEONS, ATOMS ALSO ALWAYS ARE WELL DEFINED, FIX, AND NANO UNIT SPECIFIC ENERGY QUANTITIES AND KINDS.

SOME FUNDAMENTAL PARTICLES ARE MULTI QUANTUM ENERGY PROVIDERS, THESE FUNDAMENTAL PARTICLE'S ENERGY CONTENTS ARE MORE THAN ONE KIND ENERGY IN ONE FUNDAMENTAL PARTICLE, THE BIO FRIENDLY RED

COLOR FUNDAMENTAL PARTICLE POSSESSES TWO KIND DIFFERENT THERMAL ENERGY, AND LIGHT ENERGY CONTENT. THE BIO FRIENDLY BLUE AND VIOLET PARTICLES POSSESSES TWO DIFFERENT KIND QUANTUM ELECTRIC AND LIGHT ENERGY CONTENTS, ETC.

THE RED COLOR FUNDAMENTAL PARTICLES AND INFRARED PARTICLES BOTH ARE QUANTUM THERMAL ENERGY PROVIDERS AS WELL AS QUANTUM LIGHT ENERGY DONORS. THE CONSTRUCTED ATOMS, ELECTRONS, NUCLEONS BY USING INFRARED AND RED FUNDAMENTAL PARTICLE COMPOUNDS CONSTRUCTIONS ALSO POSSESS THERMAL AND LIGHT ENERGY CONTENTS, AND ALSO ARE THERMAL AND LIGHT ENERGY PROVIDER PROTONS, NEUTRONS, ELECTRONS, ATOMS, MOLECULES AND BIOMOLECULES INSIDE THE CELLS.

THE CELL'S MOLECULAR AND BIOMOLECULAR CONSTRUCTIONS THROUGH USING THE RED PARTICLES, INFRARED PARTICLES, BLUE OR VIOLET FUNDAMENTAL PARTICLES, SONIC PARTICLES, AND MANY OTHER NONDISCOVERED THERMAL ELECTRIC, LIGHT AND SONIC FUNDAMENTAL PARTICLES OTHER THAN ABOVE, ALSO ALL ARE DIFFERENT KIND ENERGY STORAGE SYSTEMS, FOR THERMAL ENERGIES, ELECTRIC ENERGIES, SONIC AND LIGHT ENERGY. THESE BIOMOLECULAR AND MOLECULAR CONSTRUCTIONS CAN PROVIDE ELECTRIC ENERGIES, THERMAL ENERGIES, LIGHT ENERGIES OR SOUND ENERGIES FOR INTERNAL CELL CYTOPLASMIC AND NUCLEAR SYSTEMS AND ORGANS, AND THROUGH THOSE DIFFERENT KIND

ENERGIES CAN OPERATE DIFFERENT BIOLOGICAL, CHEMICAL, PHYSICAL, MECHANICAL TASKS.IN ANY GIVEN QUANTUM LOCATIONS, OR IN ANY GIVEN MICRO LOCATION INSIDE THE CELLS.

THE PARTICIPATION OF THESE THERMAL, LIGHT, ELECTRIC, PARTICLE COMPOUND MOLECULAR COMPOUNDS IN PRODUCTION OF DIFFERENT KIND ATOMS, ELECTRONS, NUCLEONS, BIOMOLECULAR COMPOUND CONSTRUCTIONS, SECONDARILY ARE PRODUCERS OF LARGER QUANTITY ENERGY DONORS.

THESE INTERA CELLULAR MICRO MOLECULAR CONSTRUCTIONS, INTERNAL CYTOPLASMIC AND INTER - NUCLEUS ORGANS AND SYSTEMS ARE ENERGY POWER HOUSES, CAN PROVIDE DIFFERENT ENERGY KINDS, IN DIFFERENT ENERGY QUANTITIES, FROM CONSTRUCTED LARGER MOLECULAR AND BIOMOLECULAR ENERGY PROVIDER UNITS TO DIFFERENT NANO SITES, MICRO LOCATIONS IN SATISFACTORY HIGH QUANTITY DIFFERENT KINDS ELECTRIC, THERMAL, LIGHT, SONIC ENERGIES, AND CAN ACHIEVE ALL KIND TASKS SUCH AS DIFFERENT BIOLOGICAL, CHEMICAL, PHYSIOLOGICAL, OR IN NON LIVE ATOMS CAN OPERATE THROUGH THOSE ENERGIES DIFFERENT PHYSICAL, INDUSTRIAL, COMMERCIAL, CHEMICAL, AND MECHANICAL TASKS.

> DEPENDING WHAT KIND PARTICLES CONSTRUCT WHAT KIND ATOMS, MOLECULAR AND BIOMOLECULAR COMPOUND CONSTRUCTIONS, THEIR ENERGY KINDS, AND ENERGY CONTENTS VARY REMARKABLY FROM ONE STRUCTURE TO THE OTHER

CONSTRUCTION. BUT ALWAYS IN ANY GIVEN MOLECULAR, BIOMOLECULAR STRUCTURES, AND CELL CONSTRUCTIONS, THE ENERGY CONTENTS, NEEDED ENERGY KINDS, AND QUANTUM ENERGY QUANTITIES ALWAYS ARE WELL DEFINED, CONSTANT, CERTAIN, FIX AND CHEMICAL STRCTURE SPECIFFIC.

ALSO DEPENDING WHAT KIND ATOMS, AND BIOMOLECULES CONSTRUCT WHAT KIND CELL'S CYTOPLASMIC OR CELL'S NUCLEAR DIFFERENT ORGANS AND SYSTEMS CONSIDERED.

IN ALL OF THOSE ABOVE CONSTRUCTIONS, THEIR NEEDED INTERNAL CELL ENERGY KINDS AND QUANTITIES (SUCH AS NEEDED ELECTRIC ENERGIES, THERMAL ENERGIES, ETC. AND WHAT ARE NEEDED AMOUNTS, IN ORDER TO ACHIEVE DIFFERENT BIOLOGICAL, PHYSIOLOGICAL, CHEMICAL FUNCTIONS, ETC.) ALL OF THESE VARIABLES, ALWAYS WILL BE REMARKABLY DIFFERENT ONE STRUCTURES SPECIES, FROM OTHER STRUCTURE OR SPECIES. BUT IN GENERAL FOR ANY GIVEN BIOMOLECULAR OR CELL SPECIES AND CONSTRUCTIONS, ALWAYS THE ENERGY KINDS, ENERGY QUANTITIES, ENERGY PRODUCTIONS, AND ENERGY USES, ALWAYS ALL ARE WELL DEFINED, FIX, CONSTANT, CELL SPECIES SPECIFIC.

ELECTRIC PARTICLE ATOMS AND MOLECULES

ALSO BETWEEN NON LIVE ELEMENTS, SOME ELECTRIC ATOMS HAVE BEEN CONSTRUCTED FROM SPECIFFIC KIND HIGHR ELECTRIC ENERGY CONTENT ELECTRIC FUNDAMENTAL PARTICLE CLASSES, WHICH THEIR FUNDAMENTAL PARTICLES POSSESSES REMARKABLY HIGHER ELECTRIC ENERGY CONTENT THAN THE REST OF THE OTHER ELECTRIC FUNDAMENTAL PARTICLE MOLECULAR COMPOUND CONSTRUCTION, AND ELECTRIC ATOM COMPOSITES.

THESE SUPER ENERGY ELECTRIC ATOMS, AND THEIR ELECTRIC COMPOUND MOLECULAR CONSTRUCTIONS AND THEIR ELECTRIC ATOM COMPOSITE ELECTRIC POWER HOUSES, ARE BETTER ELECTRIC ENERGY PROVIDERS THAN THE REMAINING ELECTRIC ATOMS AND MOLECULAR CONSTRUCTIONS. THESE SUPER ELECTRIC ENERGY PROVIDER ATOMS AND MOLECULAR STRUCTUES CAN DELIVER HIGHER ELECTRIC ENERGIES FOR NEEDED COMMERCIAL INDUSTRIES, TO BE USED TO CONSTRUCT POWERFUL HIGH ELECTRIC ENERGY BATTERIES FOR OPERATION OF DIFFERENT INDUSTRIAL, COMMERCIAL INSTRUMENTS, TOOLS, VEHICLES WHO ARE IN NEED OF HIGHER ELECTRIC ENERGIES IN DIFFERENT FIELDS.

LIGHT PARTICLE ATOMS AND MOLECULES

IN COMPARING DIFFERENT LIGHT PARTICLE ATOMS WITH EACH OTHER, SOME NON LIVE OR LIVE LIGHT FUNDAMENTAL PARTICLE COMPOUND CONSTRUCTED ATOMS AND MOLECULES, UNDER A. S. I. FP. Mol. CIC.

SOME ATOMS HAVE BEEN CONSTRUCTED FROM HIGHER LIGHT FUNDAMENTAL PARTICLE COMPOUND CONCENTRATIONS THAN THE OTHERS LIGHT PARTICLE CONSTRUCTED ATOMS (WHICH THE LATER ATOM STRUCTURES HAVE LESSER VARIATION LIGHT FUNDAMENTAL PARTICLE COMPOUND CONSTRUCTIONS).

THE EXAMPLES OF LIGHT PARTICLE ATOMS WITH HIGHER LIGHT FUNDAMENTAL PARTICLE MOLECULAR CONCENTRATIONS, AND POSSESSION OF THE NUMEROUS DIFFERENT LIGHT FUNDAMENTAL PARTICLES, IN LIGHT ATOMS DIFFERENT ELECTRONS AND NUCLEONS ARE TWO FOLLOWING LIGHTATOM MOLECULAR COMPOUNDS CONSTRUCTIONS AS:

1 – THE LIGHT PARTICLE ATOM, MOLECULAR COMPOUND CONSTRUCTION H-O-H.

2- THE CRYSTALLINE ATOMS, AND CRYSTALLINE MOLECULES, WHICH THESE ATOMS AND MOLECULES HAVE BEEN CONSTRUCTED FROM NUMEROUS DIFFERENT VARIATION LIGHT FUNDAMENTAL ARTICLE CLASSES AND

THEIR DIFFERENT LIGHT FUNDAMENTAL PARTICLE MOLECULAR COMPOUNDS CONSTRUCTIONS.

THE LIGHT FUNDAMENTAL PARTICLE COMPOUND CONSTRUCTED ATOMS, MOSTLY HAVE BEEN CONSTRUCTED FROM CHEMICAL COMBINATIONS OF THE DIFFERENT CLASSES OF DIFFERENT KIND LIGHT FUNDAMENTAL PARTICLES WITH EACH OTHER, OR THROUGH COMBINATIONS OF THOSE LIGHT PARTICLE MOLECULES WITH DIFFERENT OTHER KINDS NON LIGHT FUNDAMENTAL PARTICLES, OR OTHER CLASSES LIGHT FUNDAMENTAL PARTICLE MOLECULAR STRUCTURES, UNDER REGENERATIVE A. S. I. FP. Mol. CIC.

IN CRYSTALLINE ATOMS, THE MOLECULAR COMPOUNDS HAVE BEEN CONSTRUCTED FROM DIFFERENT CLASSES OF DIFFERENT KIND LIGHT FUNDAMENTAL PARTICLES, UNDER REGENERATIVE A. S. I. FP. Mol. CIC.

THESE CONSTRUCTING LIGHT FUNDAMENTAL PARTICLES SOME ARE SNESIBLE DIFFERENT LIGHT PARTICLE CLASSES, (SUCH AS: RED LIGHT FUNDAMENTAL PARTICLES, GREEN AND YELLOW COLOR LIGHT FUNDAMENTAL PARTICLES, OR BLUE AND VIOLET COLOR LIGHT FUNDAMENTAL PARTICLE CLASSES, ETC.).

BUT MOST OF CRYSTALLINE ATOMS LIGHT FUNDAMENTAL PARTICLE MOLECULAR CONSTRUCTIONS ARE NON DISCOVERED NON SENSIBLE LIGHT PARTICLES, ALSO THERE ARE SOME SENSIBLE BIO FRIENDLY DIFFERENT CLASSES OF

LIGHT PARTICLES, WHICH HAVE NOT BEEN DISCOVERED AT PRESENT TIME, AND PRESENTLY ARE MOSTLY UNKNOWN,

THE CRYSTALLINE ATOMS LARGER COMPLEX LIGHT PARTICLE COMPOUNDS CONSTRUCTIONS, THROUGH DEGENERATIVE A. S. I. FP. Mol. CIC. BREAK DOWN INTO SMALLER CONSTRUCTING LIGHT FUNDAMENTAL PARTICLES, AND CAN BE VISUALIZED AS FREE AIRBORNE RED, YELLOW, GREEN, BLUE, VIOLET, ETC. COLOR LIGHT FUNDAMENTAL PARTICLES AT SPACE. DEFINITELY NONSENSIBLE LIGHT PARTICLE CLASSES ARE NOT DISCOVERED YET, AND CAN NOT BE IDENTIFIED PRESENTLY.

THE GREEN PARTICLES AND YELLOW PARTICLES BOTH POSSESSES LIGHT ENERGY CONTENTS.

THE INFRARED, AND RED PARICLES BOTH POSSESSES QUANTUM THERMAL AND LIGHT ENERGY CONTENT. THE VIOLET PARTICLES, BLUE PARTICLES POSSESSES QUANTUM ELECTRIC ENERGY, AND LIGHT ENERGY CONTENTS. ALL OF ABOVE PARTICLES CAN PROVIDE ENERGIES INTO DIFFERENT NANO LOCATIONS. CAN PERFORM QUANTUM NANO TASKS, INSIDE ELECTRONS, NEUTRONS, PROTONS AND ATOMS.

IN NON LIVE INDUSTRIAL WORLD, THE PRODUCED LIGHT PARTICLE CURRENTS CAN BE USED IN ELECTRONIC, AND DIGITAL INDUSTRIES, PHOTO INDUSTRIES, COMMERCIAL, INDUSTRIAL FIELDS, TRAFFIC, ADVERTISEMENTS, ETC. OR ART FIELDS AND CREATIONS.

THE BIOLOGICAL USE OF LIGHT, THERMAL, ELECTRIC, SONIC FUNDAMENTAL PARTICLES HAVE BEEN EXPLAINED IN

OTHER CHAPTERS, AS WELL IN ANOTHER BOOKS OF THE AUTHOR.

COLD FUNDAMENTAL PARTICLE CLASSES

MOST OF THESE FUNDAMENTAL PARTICLES ARE BIO LETHAL, ONLY SMALL GROUPS OF COLD FUNDAMENTAL PARTICLE CLASSES IN EARTH ARE BIO FRIENDLY PARTICLES.

Construction of Nano Unit Creation of Electrons, Neutrons, Protons

---

MOST ELECTRONS, NEUTRONS, PROTONS, AND ATOMS HAVE BEEN CONSTRUCTED FROM NUMEROUS DIFFERENT KIND THERMAL- ELECTRIC- LIGHT- SONIC- ETC. PARTICLE MOLECULAR COMPOUND CONSTRUCTIONS, UNDER THE REGENERATIVE AND DEGENERATIVE A. S. I. FP. Mol. CIC. MOST NANO UNITS ARE STORAGE SYSTEMS OF DIFFERENT

QUANTUM NEEDED ENERGIES, ALSO STORE AND RELEASE FREE NEEDED FUNDAMENTAL PARTICLES.

NANO UNIT'S PARTICLE INTELLIGENCE SYSTEM CENTERS (PIS) OPERATE ELECTRONS, NUCLEONS QUANTUM FUNCTIONS, THE ENERGY STORAGE AND RETRIEVAL SYSTEMS, REGENERATIVE AND DEGENERATIVE A. S. FP. Mol. CIC. AND PHYSICAL, CHEMICAL, BIOLOGICAL FUNCTIONS OF NANO UNITS.

INTERNAL ELECTRON, NEUTRON, PROTON PARTICLE CIRCULATION SYSTEMS (PCS) ARE BI – DIRECTIONAL FUNDAMENTAL PARTICLE TRANSIT CURRENTS AND FUNDAMENTAL PARTICLE CIRCULATION SYSTEMS, WHICH THE AFFERENT PCS RUNS, IN OPPOSITE DIRECTIONS OF THE EFFERENT PCS. ONE GROUP PCS CURRENTS BRING INTO ATOM THE NEEDED PARTICLE CLOUDS AND FUNDAMENTAL PARTICLES, AND THE OPPOSITE DIRECTION PCS CURRENTS REMOVE THE EXCESS NOT NEEDED FUNDAMENTAL PARTICLES FROM INSIDE ELECTRONS AND NUCLEONS AND CARRY NON NEEDED PARTICLES TO THE OUTSIDE ELECTRONS, NUCLEONS, AND ATOMS.

Homogenous Fundamental Particle construction Electrons, Neutrons, Protons, Atoms

THE ATOMS, THAT HAVE BEEN CONSTRUCTED FROM SPECIFIC FUNDAMENTAL PARTICLES

THERE ARE ALSO ANOTHER GROUP NANO UNITS, WHICH HAVE BEEN CONSTRUCTED ONLY FROM A FEW KIND DIFFERENT PARTICLE MOLECULAR COMPOUND CONSTRUCTIONS.

FOR EXAMPLE, THE ELECTRIC FUNDAMENTAL PARTICLE COMPOUND CONSTRUCTED ELECTRONS, NUCLEONS, ATOMS HAVE BEEN CONSTRUCTED FROM ELECTRIC FUNDAMENTAL PARTICLE COMPOUND CONSTRUCTIONS. THERMAL FUNDAMENTAL PARTICLE COMPOUND CONSTRUCTION ELECTRONS, NEUTRONS, PROTONS PRODUCE THERMAL ATOMS.

IN SOME CASES OF THERMAL PARTICLE CONSTRUCTED ELECTRONS, NUCLEONS, OR ELECTRIC FUNDAMENTAL PARTICLE COMPOUND CONSTRUCTED ATOMS, THE THERMAL OR ELECTRIC ENERGY CONTENTS OF THEIR ELECTRONS, NEUTRONS, PROTONS, ARE THOUSANDS

FOLDS MUCH MORE THAN THE ORDINARY ATOM'S ELECTRIC OR THERMAL ENERGY QUANTITIES.

THE ELECTRIC FUNDAMENTAL PARTICLE COMPOUND CONSTRUCTION ELEMENTS, ARE SUITABLE FOR CONSTRUCTION OF QUANTUM BATTERIES, THEY CAN PROVIDE REMARKABLE ELECTRIC ENERGY FOR COMMERCIAL AND INDUSTRIAL QUANTUM BATTERIES, MUCH BETTER THAN ORDINARY MOLECULAR FUNDAMENTAL PARTICLE CONSTRUCTED COMPOUNDS. ABOVE FACTS ARE TRUE IN REGARD TO THERMAL ATOMS, SONIC ATOMS, ETC. AS WELL.

SPECIFIC ENERGY CONTENT ELECTRONS, PROTONS, NEUTRONS, ATOMS

THERMAL FUNDAMENTAL PARTICLE COMPOUND CONSTRUCTED ELECTRONS, NUCLEONS, ATOMS ARE THERMAL ENERGY STORAGE SYSTEMS, THESE THERMAL ENERGY NANO UNITS, OR MICRO UNITS SYSTEMS, CAN PROVIDE THERMAL ENERGY FOR NEEDED QUANTUM LOCATIONS, OR MICRO

LOCATIONS, ACCORDING SPECIFIC NEEDED AND CALIBERATED THERMAL DOSES.

THE ABOVE FACTS ALSO ARE TRUE IN REGARD TO ALL OTHER SPECIFIC FUINDAMENTAL PARTICLE CONSTRUCTED ATOMS, ELECTRONS, NUCLEONS. SUCH AS ELECTRIC, GAMMA, SONIC, LIGHT, OR X- FUNDAMENTAL PARTICLE COMPOUND CONSTRUCTION ATOMS, NEUTRONS, PROTONS, AND ELECTRONS.

THE BIO LETHAL THERMAL FUNDAMENTAL PARTICLES THAT CONSTRUCTING THE MAGMA ELECTRONS, NUCLEONS FUNDAMENTAL PARTICLE COMPOUNDS CONSTRUCTIONS, ARE DIFFERENT THAN THE BIO FRIENDLY THERMAL FINDAMENTAL PARTICLES, THAT CONSTRUCT LIVING THE THINGS ATOMS, ELECTRONS, NEUTRONS, PROTONS THERMAL FUNDAMENTAL PARTICLE COMPOUND CONSTRUCTIONS.

THE THERMAL ATOMS ARE DIFFERENT THAN THE LIGHT FUNDAMENTAL PARTICLE CONSTRUCTION ATOMS AND NANO UNITS, THE ABOVE RULES ARE TRUE IN REGARD TO ALL OTHER DIFFERENT FUNDAMENTAL PARTICLE CLASSES AS WELL.

CRYSTALLINE ATOM IS ONE EXAMPLE FROM LIGHT PARTICLE CONSTRUCTION ATOMS, AND CRYSTALLINE IS LIGHT ENERGY PROVIDER AND LIGHT FUNDAMENTAL PARTICLE DONOR ALSO.

THE THERMAL PARTICLE COMPOUND CONSTRUCTION ATOMS CAN PROVIDE QUANTUM THERMAL ENERGY

TO NANO LOCATIONS, MUCH BETTER THAN LIGHT FUNDAMENTAL PARTICLE COMPOUND CONSTRUCTED ATOMS. AT THE SAME TIME THE LIGHT FUNDAMENTAL PARTICLE CONSTRUCTED ATOMS CAN PROVIDE DIFFERENT KINDS LIGHT FUNDAMENTAL PARTICLES FOR MEDIA AS WELL AS THEY CAN PROVIDE LIGHT ENERY TO ANY GIVEN QUANTUM SITE, OR MACRO LOCATIONS.

THE THERMAL ENERGY PROVIDER ATOMS OR ANY OTHER TYPE ENERGY CONSTRUCTED ATOMS, CAN PROVIDE THERMAL FUNDAMENTAL PARTICLE MOLECULES, OR ANY OTHER KIND FUNDAMENTAL PARTICLES, OR ENERGY KINDS TO BE USED FOR DIFFERENT QUANTUM LOCATIONS FOR TRIGGERING DIFFERENT A. S. I. FP. Mol. CIC. OR ANY OTHER KIND DIFFERENT CHEMICAL, BIOLOGICAL, PHYSICAL, MECHANICAL USES.

LIGHT PARTICLE CONSTRUCTED ATOMS CAN PROVIDE DIFFERENT KIND LIGHT FUNDAMENTAL PARTICLES FOR CHEMICAL INTERACTIONS, AS WELL AS LIGHT ATOMS UNDER DEGENERATIVE A. S. I. FP. Mol. CIC. CAN PRODUCE DIFFERENT COLOR LIGHT FUNDAMENTAL PARTICLE CLOUDS FOR COMMERCIAL AND INDUSTRIAL USE. AND LIGHT ATOM PROVIDED DIFFERENT TYPES LIGHT ENERGIES, AND DIFFERENT COLOR PRODUCED LIGHT FUNDAMENTAL PARTICLES, ALL CAN USED IN DIFFERENT COMMERCIAL AND INDUSTRIAL FIELDS.

STORAGE OF FUNDAMENTAL PARTICLE INSIDE ATOMS UNDER REGENERATIVE A. S. I. FP. Mol. CIC, AS WELL AS

RETRIEVAL OF DIFFERENT FUNDAMENTAL PARTICLE FROM INSIDE ELECTRONS, NUCLEONS TO THE OUTSIDE, UNDER DEGENERATIVE A. S. I. FP. Mol. CIC. APPLY FOR ALL DIFFERENT KIND EXISTING UNIVERSAL BIO FRIENDLY AND BIO HOSTILE FUNDAMENTAL PARTICLES WITH NO EXCEPTIONS.

QUANTUM ENERGY STORAGE AND QUANTUM ENERGY RETRIEVAL IN QUANTUM LOCATIONS

FUNDAMENTAL PARTICLES NATURALLY POSSESS INHERITANT FUNDAMENTAL PARTICLE SPECIFIC ENERGIES, WEIGHT AND GRAVITON. FUNDAMENTAL PARTICLES TRANSIT THROUGH DIFFERENT PARTICLE TRANSMISSION ROUTES INTO INSIDE ELECTRONS NUCLEONS, AND UNDER REGENERATIVE A. S. I. F.P. Mol. C.I.C. COMBINE WITH INTER ELECTRON NUCLEON PARTICLE COMPOUNDS AND CONSTRUCT ELECTRONS NUCLEONS PARTICLE- COMPOUND CONSTRUCTIONS, THEREFORE FUNDAMENTAL PARTICLES STORE THEIR ENERGY CONTENT IN PARTICLE COMPOUND FORMS INSIDE ELECTRONS NUCLEONS, THIS IS PHENOMENON OF STORAGE OF ENERGY INSIDE ELECTRONS NUCLEONS.

FUNDAMENTAL PARTICLES CAN TRIGGER START, ACHIEVE AND CONTROL QUANTUM CHEMICAL, PHYSICAL, BIOLOGICAL TASKS, INCLUDING A. S. I. F.P. Mol. C.I.C. INSIDE ANY GIVEN QUANTUM LOCATION THROUGH PROVIDING,

OR STOPPING THE SOURCES OF NEEDED ENERGIES FOR TASKS.

PRESENCES OF ANY FUNDAMENTAL PARTICLES IN ANY QUANTUM LOCATIONS, IS EQUAL TO EXISTENCES OF FUNDAMENTAL PARTICLE ENERGY CONTENT, IN THAT GIVEN QUANTUM LOCATION.

UNDER DEGENERATIVE A. S. I. F.P. Mol. C.I.C. THE CONSTRUCTED LARGE MOLECULAR FUNDAMENTAL PARTICLES COMPOUNDS BREAK DOWN INTO SMALLER CONSTRUCTING FREE PARTICLES FORMS, RELEASE FUNDAMENTAL PARTICLES, AND ENERGY INTO QUANTUM LOCATION AS ENERGY DONERS, (PHENOMENON OF RETRIEVAL OF FUNDAMENTAL PARTICLE S TO OUTSIDE ELECTRONS, NUCLEONS), AND ACHIEVE NEEDED QUANTUM FUNCTIONS, INSIDE OR OUTSIDE DIFFERENT ELECTRONS, NUCLEONS ALL OVER UNIVERSE UNDER THE MATTER'S PHYSIC, CHEMISTRY LAWS, AND RULES. FUNDAMENTAL PARTICLES PROVIDE OR START ENERGY IN ORDER TO START OR STOP CHEMICAL, BIOLOGICAL, PHYSICAL, MECHANICAL TASKS AND INTERACTIONS, INSIDE NANO LOCATIONS. PARTICLES ACHIEVE, CONTROL AUTONOMOUS SEQUENTIAL PHYSICAL CHEMICAL INTERACTIONS, TASKS, AND ALL KIND QUANTUM FUNCTIONS, IN ANY GIVEN NANO LOCATION.

Biochemistry of Fundamental Particles COLOR OF PLANTS

{(Electron, Nucleon, Atom, Biomolecule, Cell, Plant, Flower) - (Genesis)}, Under pathways of World's Molecular Evolution in each spring. under A. S. I. FP. Mol. CIC. through Molecular Evolution Paths, Orders and Laws.

Color Genesis &Tissue Genesis under Evolution
PLANT GENESIS

SUN ORIGIN FUNDAMENTAL PARTICLES MOSTLY ARE NON SENSIBLE AND NON DISCOVERED, SOME SENSIBLE SUN ORIGIN LIGHT PARTICLES ALSO ARE NOT DISCOVERED YET, THIS CLASS LIGHT PARTICLES UNDER DEGENERATIVE A. S. I. FP. Mol. CIC CAN BE FREED FROM COMPOUND MOLECULAR PARTICLE STRUCTURES INTO AIR. EARTHS DISCOVERED LIGHT FUNDAMENTAL PARTICLES ARE SUCH AS RED, YELLOW, GREEN, BLUE, VIOLET AND OTHER FUNDAMENTAL PARTICLE CLASSES.

FOR EXAMPLE, RED COLOR, BLUE, GREEN, OR VIOLET, YELLOW FUNDAMENTAL PARTICLES DIRECTLY FROM AIR UNDER PLANT ATOM'S ATTRACTIVE GRAVITON FORCES, GET ATTRACTION INTO ATOMS, THESE LIGHT PARTICLES THROUGH INTERNAL ELECTRON AND NUCLEON PARTICLE CIRCULATION SYSTEMS (PCS). TRAVEL INTO INSIDE ATOMS SUBSYSTEM UNIT CHEMICAL LAB.S, COMBINE WITH INTERNAL ELECTRON NUCLEON PARTICLE COMPOUNDS, UNDER A. S. I. FP. Mol. CIC. THE MOLECULAR COMBINATION PRODUCE RED, YELLOW, GREEN, BLUE, VIOLET OR OTHER MIXED COLORS DIFFERENT TYLES LIGHT FUNDAMENTAL

PARTICLE COMPOUND CONSTRUCTIONS, INSIDE FLOWER'S ELECTRONS, NUCLEONS.

THIS IS PHENOMENON OF NEO ELECTRONS GENESIS, NEO ATOM GENESIS, NEO NEUTRON- PROTON GENESIS. AND GENESIS OF NEW COLOR AND CONSTRUCTION CELLS FOR PLANT SPECIES.

RED COLOR ATOMS TINT RED COLOR CELLS AND FLOWERS. GREEN COLOR LIGHT PARTICLES CONSTRUCT GREEN COLOR LEAVES PARTICLE COMPOUNDS, PRODUCE GREEN ATOMS AND CELLS. YELLOW, VIOLET, BLUE COLOR LIGHT PARTICLE COMPOUND PRODUCTIONS PRODUCE THE SAME COLOR DIFFERENT FOLOWER SPECIES. THE MIXED LIGHT PARTICLE COMBINATIONS WITH INTERNAL ELECTRON, NUCLEON PARTICLE COMPOUNDS, PRODUCE ENORMOUS DIFFERENT OTHER COLOR FLOWERS.

ALL OF THESE A. S. I. FP. Mol. CIC. AND GENESIS OF DIFFERENT SPECIES PLANTS IN EACH SPRING, ALL ARE UNDER EXACT LAWS, ORDERS, AND PATHS OF PARTICLE MOLECULAR ELVOLUTIONS OF THE PAST WORLD HISTORY EXACTLY SAME.

THE BIOMOLECULAR AUTONOMOUS SEQUENTIAL CHEMICAL INTERACTION CYCLES AND CELL MUTATIONS ALSO FOLLOW PAST WORLD HISTORY'S CELL AND BIOMOLECULAR EVOLUTION PATHS, LAWS, ORDERS. UNDER THE CELL, AND BIOMOLECULAR EVOLUTION FOOT STEPS OF THE PAST, UNDER FUNDAMENTAL PARTICLE PRECISION ACCURACY.

NON DISCOVERED NON SENSIBLE BIO FRIENDLY ELECTRIC FUNDAMENTAL PARTICLES IN EARTH

(ND. NS. BF. E. FP)

==

ELECTRIC & LIGHT ENERGY PROVIDERS TO PLANT'S INTERNAL ELECTRONS, NUCLEON'S NANO LOCATIONS AND MICRO SITES.

VIOLET COLOR ELECTRIC PARTICLES (V. E.– FP), BLUE COLOR ELECTRIC PARTICLES (B. E. – FP)

BLUE COLOR LIGHT PARTICLES (B. Y. - FP), VIOLET COLOR LIGHT PARTICLE (V. Y. – FP)

V. E.-FP = V. Y. – FP   B. E. – FP = B. Y. – FP

THE BIO FRIENDLY, SENSIBLE VIOLET COLOR FUNDAMENTAL PARTICLES (V – FP) AND BLUE COLOR FUNDAMENTAL PARTICLES (B – FP), AS WELL AS ALL ND. NS. BF. - E. Y. - FP ALL IN FUNDAMENTAL PARTICLE CONTENTS INHERITANCE POSSESSES MIXED KIND LIGHT ENERGY CONTENT AND ELECTRIC ENERGY CONTENTS. THE "V –FP, B –FP AND ND. NS. BF. – E. Y. – FP" FUNDAMENTAL PARTICLES ARE LIGHT ENERGY PROVIDER FUNDAMENTAL PARTICLES AS WELL AS ARE ELECTRIC ENERGY PROVIDER FUNDAMENTAL PARTICLES, WHICH THEY CAN DONATE THEIR LIGHT OR ELECTRIC ENERGY CONTENTS INTO INTERNAL ELECTRON, NUCLEON NANO LOCATIONS AND MICRO LOCATIONS. WHICH THESE DIFFERENT ENERGY TYPES ARE UISED TO PERFORM NANO TASKS AND MICRO TASKS IN NANO SITES AND MICRO LOCATIONS.

IN LIVING THINGS THESE PARTICLES PROVIDE ELECTRIC ENERGIES AND LIGHT ENERGIES INTO ELECTRON NUCLEON QUATUM LOCATIONS, WITH COOPERATION OF OTHER BIO FRIENDLY FUNDAMENTAL PARTICLES OPERATE ENTIRE LIVING THINGS BIOLOGICAL, CHEMICAL, PHYSICAL NANO FUNCTIONS IN DIFFERENT QUANTUM LOCATIONS OF PLANTS AND ANIMAL SPECIES.

Plant quantum energy providers for achieving Nano Tasks and Micro tasks in quantum locations Inside Atoms, Cells. Under: A. S. I. FP. Mol. CIC.

RED COLOR PARTICLE'S ENERGY CONTENTS MOSTLY ARE THERMAL ENERGY AS WELL AS HAS LIGHT ENERGY.

BLUE COLOR PARTICLES, VIOLET COLOR PARTICLES ENERGY CONTENTS ARE ELECTRIC ENERGIES, ALSO LIGHT ENERGY AS WELL, THEREFORE, THE RED FUNDAMENTAL PARTICLE COMPOUND MOLECULAR CONSTRUCTIONS INSIDE PLANT'S ELECTRONS, NUCLEONS, WHICH HAS BEEN PRODUCED UNDER REGENERATIVE AUTONOMOUS SEQUENTIAL INTER FUNDAMENTAL PARTICLE CHEMICAL INTERACTIONS CYCLES, AND CHAINS (REG. A. S. I. F.P. Mol. C.I.C.). POSSESSES LIGHT ENERGY STORAGE SYSTEMS, AND THERMAL ENERGY STORAGE SYSTEMS.

THE NON DISCOVERED, NON SENSIBLE, BIO FRIENDLY LIGHT FUNDAMENTAL PARTICLE CLASSES (ND- NS – BF – Y. FP) PARTICLE COMPOUND MOLECULAR CONSTRUCTIONS, AS WELL AS THE BLUE AND VIOLET PARTICLE COMPOUND

MOLECULAR CONSTRUCTIONS, ALL STORE BIO FRIENDLY ELECTRIC ENERGIES INSIDE PLANT'S ELECTRONS, NUCLEONS, ATOMS, BIOMOLECULAR STRUCTURES AND CELLS.

THE PRODUCED RED PARTICLE COMPOUND CONSTRUCTIONS, OR ABOVE MENTIONED ANY OTHER FUNDAMENTAL PARTICLE COMPOUNDS INSIDE PLANT'S INTERNAL ELECTRONS, AND NUCLEONS. WHEN THESE LARGE MOLECULES BREAK DOWN INTO SMALLER PARTICLE MOLECULAR CONSTRUCTIONS (INTO NON -COMPOUND FREE FUNDAMENTAL PARTICLE MOLECULAR STATUS) SUCH AS FREE RED PARTICLE MOLECULES, OR FREE BLUE, GREEN, VIOLET, YELLOW PARTICLE MOLECULAR STATUS, ETC. UNDER THE DEGENERATIVE A. S. I. FP. Mol. CIC.

IN REGARD TO RELEASED ABOVE MENTIONED DIFFERENT FREE RED, YELLOW, VIOLET, BLUE, GREEN FUNDAMENTAL PARTICLES IN INTERNAL ELECTRON NUCLEON QUANTUM LOCATIONS, PROVIDE FREE RED, BLUE, GREEN, OR VIOLET AND YELLOW FUNDAMENTAL PARTICLE, READY TO BE USED IN ANOTHER CHAINS OF AUTONOMOUS SEQUENTIAL CHEMICAL INTERACTIONS (REGENERATIVE A. S. I. FP. Mol. CIC.).

ALSO THE FREED ENERGIES FROM RED, GREEN, YELLOW, BLUE, VIOLET PARTICLES (IN FORMS OF THERMAL, LIGHT, ELECRIC, SONIC ENERGY FORMS, ETC.), ALSO THESE DIFFERENT KIND DIFFERENT ENERGIES CAN BE USED IN DIFFERENT QUANTUM TASKS, WHICH ARE IN NEEDS OF DIFFERENT KIND ENERGIES. IN DIFFERENT INTERNAL ELECTRON NUCLEON QUANTUM LOCATIONS, IN ORDER TO

CONDUCT AND ACHIEVE DIFFERENT NEEDED QUANTUM BIOLOGICAL, CHEMICAL, PHYSICAL, MECHANICAL FUNCTIONS INSIDE DIFFERENT PLANT'S INTERNAL ELECTRONS, NUCLEONS, ATOMS, BIOMOLECULES, AND CELLS LOCATIONS. AS WELL AS TO ACHIEVE NEEDED MICRO TASKS INSIDE BIOMOLECULES, AND CELL CONSTRUCTIONS. ABOVE IS ENERGY STORAGE AND RETRIEVAL PHENOMENONS, INSIDE ATOMS, CELLS OF PLANTS AND OTHER LIVING THINGS.

SENSIBLE RED YELLOW, GREEN, BLUE, VIOLET FUNDAMENTAL PARTICLES STORAGE AND RETRIEVAL SYSTEMS APPLY FOR ALL OTHER DIFFERENT CLASSES OF OTHER KIND FUNDAMENTAL PARTICLES ACROSS THE UNIVERSE FOR ALL OTHER FUNDAMENTAL PARTICLES AS WELL.

GREEN COLOR LIGHT FUNDAMENTAL PARTICLES (GY – FP) A. S. I. FP. Mol. CIC. IN PLANTS

LIGHT ENERGY POVIDER FUNDAMENTAL PARTICLES

UNIT OF LIGHT ENERGY

THE EXISTING QUANTUM LIGHT ENERGY QUANTITY CONTENT, IN ONE GREEN LIGHT FUNDAMENTAL PARTICLE IS EQUAL TO ONE UNIT LIGHT ENERGY.

THE GREEN LIGHT FUNDAMENTAL PARTICLE (GY – FP) IS AN ESSENTIAL BIO FRIENDLY DOMINANT LIGHT FUNDAMENTAL PARTICLES IN CONSTRUCTION OF PLANT'S AND LIVING THINGS INTERNAL ELECTRON, NUCLEON PARTICLE COMPOUND CONSTRUCTION, IN EARHT PLANET.

THE LIGHT ENERGY AMOUNT OF ONE GREEN COLOR LIGHT FUNDAMENTAL PARTICLE (GY –FP) IN PLANET EARTH, IS EQUAL TO ONE UNIT LIGHT ENERGY AMOUNT (UYE). THIS UNIT IS USED IN MEASUREMENTS OF QUANTUM QUANTITIES LIGHT ENERGIES.

THE DIFFERENT LIGHT FUNDAMENTAL PARTICLE CLASSES HAVE DIFFERENT QUANTITY QUANTUM LIGHT ENERGY CONTENTS. THE EXISTING QUANTUM LIGHT ENERGY CONTENT QUANTITIES OF OTHER DIFFERENT FUNDAMENTAL PARTICLES ENERGIES, CAN BE QUANTIZED, COMPARED AND MEASURED IN ACCORDANCES AND COMPARISONS TO THE LIGHT ENERGY QUANTITY OF ONE GREEN LIGHT FUNDAMENTAL PARTICLE IN PLANET EARTH. WHICH IT IS EQUAL TO ONE UNIT LIGHT ENERGY.

THE YELLOW COLOR LIGHT FUNDAMENTAL PARTICLE (Y. y.- FP) IN EARTH IS ANOTHER EXAMPLE FROM SENSIBLE BIOFRIENDLY LIGHT PARTLE CLASSES, THERE ARE LARGE NUMBERS OF OTHER NON DISCOVERED AND NON SENSIBLE

LIGHT FUNDAMENTAL PARTICLE CLASSES IN EARTH WHICH HAVE BEEN USED IN CONSTRUCTIONS OF DIFFERENT NANO UNITS.

UNIT OF ELECTRIC ENERGY

THE EXISTING QUANTUM ELECTRIC ENERGY QUANTITY CONTENT, IN ONE ULTRAVIOLET ELECTRIC FUNDAMENTAL PARTICLE, IS EQUALTO ONE QUANTUM UNIT OF ELECTRIC ENERGY.

AN ULTRAVIOLET FUNDAMENTAL PARTICLE (UV. E – FP) IS WEAKEST KNOWN ELECTRIC FUNDAMENTAL PARTICLE IN EARTH, BETWEEN THE ALL OTHER KNOWN EXISTING OTHER ELECTRIC FUNDAMENTAL PARTICLE CLASSES, WHICH ALMOST ALL ARE BIO LETHAL FUNDAMENTAL PARTICLES. THE ELECTRIC ULTRAVIOLET FUNDAMENTAL PARTICLE POSSESSES LEAST PARTICLE WEIGHT, LEAST PARTICLE GRAVITON FORCE AND LEAST QUANTUM QUANTITY ELECTRIC ENERGY CONTENT, IN COMPARISON TO THE OTHER EARTH'S INDUSTRIAL ELECTRIC FUNDAMENTAL PARTICLES.

THE ELECTRIC ENERGY AMOUNT OF ONE ULTRAVIOLET FUNDAMENTAL PARTICLE IS EQUAL TO ONE QUANTUM UNIT ELECTRIC ENERGY QUANTITY, THE ULTRAVIOLET PARTICLE IS LOCATED AFTER FINISH LINES OF LIGHT FUNDAMENTAL PARTICLE CLASS, AND JUST AT STRATING

POINTS OF HARSH BIOHOSTILE HIGH ENERGY INDUSTRIAL ELECTRIC FUNDAMENTAL PARTICLE CLASSES.

ESSENTIAL RULES AND LAWS OF GENERAL FUNDAMENTAL PARTICLE CHEMISTRY
UNDER A. S. I. FP. Mol. CIC.

In Chemical Interaction between Fundamental Particle Molecules, always one Normal Molecule from one Molecular Construction combine with one Normal Molecule from another Particle Molecular Construction, and produce Tertiary other kind one normal Particle Molecule Compound. And Complete a given chain from a Given A. S. I. FP. Mol. CIC.

During the Fundamental Particle Chemical Combinations, the Dominant Fundamental Particle Structures always by having higher more powerful attractive Graviton Forces stay in center and construct Nucleus Position of Molecule, and the Recessive Fundamental Particle through weaker Graviton Force stay in periphery, and construct Peripheral Molecular Structures.

In order to Complete a given Chemical Interaction Combination Chain thoroughly between two different combining Particle Molecules. Always we need the total fundamental particle population numbers of two combining particle molecules relatively be equal, in order to complete one chain of A. S. I. FP. Mol. CIC. thoroughly and completely.

In Chemical Combinations between the Exogenous and Indigenous Particle Molecules, mostly The Lesser Particle Weight, lesser Particle Energy quantity, and lesser Graviton Force Exogenous (external atom) Fundamental Particle Molecules combine with weaker less graviton force, lesser energy content, and lesser molecular weight internal Electron particle compounds (the Indigenous internal electron, nucleon Particle Compounds), under the A. S. I. FP. Mol. CIC. and produce lesser weight, lesser energy content, lesser graviton force Tertiary Particle Molecular Compounds, and construct the internal Electron particle compound constructions with lesser Molecular weight particle compound constructions. THIS IS PHENOMENON OF ELECTRON NEO GENESIS, WITH LESSER WEIGHT, LESSER GRAVITON, LESSER ENERGY CONTENT PARTICLE MOLECULAR COMPOUNDS. THIS IS ALSO NEO PARTICLE MOLECULAR COMPOUND GENESIS OF THE NANO UNITS.

In Contrast, in Chemical combinations between the Exogenous and Indigenous Fundamental Particle Molecules, The Higher Energy Content, higher Graviton Force, and higher Particle Weights Exogenous (External Atom) Particle Molecules mostly Combine with high Energy Content, heavy Weight, higher Graviton Force internal Neutron, internal Proton Indigenous (internal Proton and internal Neutron) Particle Molecular Compounds Structures and produce (Construct) the Heavy Weight, More Graviton Force internal Nucleon's Particle Compound Molecular Constructions (Heavy Molecular compound weight construction), under the A. S. I. FP. Mol. CIC.

THIS IS PHENOMENON OF NEO PARTICLE COMPOUND GENESIS, NEO PROTON GENESIS, AND NEO NEUTRON GENESIS.

In Chemical Combinations between Exogenous and Indigenous Particle Molecular Compounds, there are enormous different other Intermediate Chemical Interaction groups, which their Exogenous and Indigenous Molecular Fundamental Particles, which are enormously different kind Fundamental particles, with different Particle Weights, different Particle Graviton Force quantities, and different quantum Particle Energy quantities, which combining these intermediate classes of different fundamental particles with each other under enormous different A. S. I. FP. M. CIC. produces enormous different Kind Construction Electrons, Neutrons, and Proton constructions. Different Planets Nano Units Constructions are remarkably different from each other.

GENERAL CHEMISTRY OF FUNDAMENTAL PARTICLES

FUNDAMENTAL PARTICLE GENERAL CHEMISTRY

Chemical combination of different Exogenous Thermal Fundamental Particles with different Internal Atom (Indigenous) Particle Compounds of a given Element

CHEMICAL COMBINATION OF EX. –FP WITH IND. – FP – COMPOUNDS

IN FOLLOWING NUMEROUS A. S. I. FP. Mol. CIC. A GIVEN CHEMICAL FORMULARY EXOGENOUS THERMAL FUNDAMENTAL PARTICLE WITH SPECIFIC PARTICLE WEIGHT, PARTICLE ENERGY, AND GRAVITON FORCE COMBINED WITH ANOTHER WELL DEFINED COMPATIBLE ONE NORMAL MOLECULE WEIGHT, ENERGY, AND GRAVITON FORCE INDIGENOUS INTERNAL ELECTRON, NUCLEON PARTICLE COMPOUND. AND THE CHEMICAL COMBINATION PRODUCED, A TERTIARY WELL DEFINED CHEMICAL FORMULARY INTERNAL ELECTRON NUCLEON INDIGENOUS NEW OTHER STRUCTURE THERMAL PARTICLE COMPOUNDS UNDER A. S. I. FP. Mol. CIC. AND EACH CHEMICAL INTERACTION ADDITIONALLY PRODUCED FREE DIFFERENT KIND PARTICLES, WHICH SCAPE THE COMBINATION FIELD.

ESSENTIAL LAWS AND RULES OF A. S. I. FP. Mol. CIC.

UNDER REGENERATIVE OR DEGENERATIVE A. S. I. FP. Mol. CIC. WHEN A GIVEN CHEMICAL COMBINATIONS TAKE PLACE BETWEEN TWO DIFFERENT FUNDAMENTAL PARTICLE MOLECULES (SUCH AS: COMBINING A GIVEN EXOGENOUS FUNDAMENTAL PARTICLE WITH ANOTHER GIVEN INDIGENOUS FUNDAMENTAL PARTICLE COMPOUNDS), ALWAYS THE TOTAL FUNDAMENTAL PARTICLE MOLECULE POPULATIONS OF BOTH COMBINING FUNDAMENTAL PARTICLE MOLECULES ARE EQUAL WITH EACH OTHER IN ORDER TO COMPLETE THE THOROUGH CHEMICAL COMBINATIONS. ALWAYS IN ALL CHEMICAL COMBINATIONS ONLY ONE MOLECULE FROM ONE GIVEN PARTICLE COMBINES WITH ANOTHER ONE MOLECULE FUNDAMENTAL PARTICLE AND COMPLETE CHEMICAL COMBINATIONS ENTIRELY ONE MOLECULE TO OTHER BETWEEN ENTIRE COMBINING MOLECULES WITH EXPONENTIAL SPEEDS OF REGENERATIVE AND DEGENERATIVE A. S. I. FP. Mol. CIC. IN MOST CASES A. S. I. FP. Mol. CIC. ARE REVERSIBLE IN OPPOSITE DIRECTIONS TO ORIGINAL PATTERNS AUTONOMOUSLY AND SEQUENTIALLY.
IN EACH GIVEN A. S. I. FP. Mol. CIC. TEST ALWAYS A WELL DEFINED CHEMICAL FORMULARY EXOGENOUS THERMAL FUNDAMENTAL PARTICLES, WHICH HAVE WELL DEFINED SPECIFIC GIVEN FUNDAMENTAL PARTICLE WEIGHT, FUNDAMENTAL PARTICLE GRAVITON FORCE AND FUNDAMENTAL PARTICLE THERMAL ENERGY QUANTITY ENTER INTO INTERNAL ELECTRON NUCLEON SUBSYSTEM UNITS CHEMICAL LAB.S THROUGH INTERNAL ELLECTRON NUCLEON PARTICLE CIRCULATION SYSTEMS (PCS) AND

COMBINE WITH INTERNAL ELECTRON NUCLEON PARTICLE COMPOUNDS.

IN DIFFERENT A. S. I. FP. Mol. CIC. TESTS, THE EX. FP. MOLECULES ALWAYS COMBINE WITH COMPATIBLE WELL DEFINED INTERNAL ELECTRON, INTERNAL NUCLEON INDIGENOUS FUNDAMENTAL PARTICLE COMPOUND CONSTRUCTIONS, WHICH THESE INDIGENOUS MOLECULAR COMPOUNDS ALL POSSESSES COMPATIBLE PARTICLE MOLECULE WEIGHT, PARTICLE GRAVITON FORCE, AND PARTICLE COMPOUND ENERGY QUANTITY CONTENTS.

FINDALLY THE EXO. FUNDAMENTAL PARTICLES INSIDE SUBSYSTEM UNITS CHEMICAL LAB.S OF RECIPIENT ATOMS UNDER A. S. I. FP. Mol. CIC. COMBINE WITH INDIGENOUS PARTICLE COMPOUNDS AND PRODUCE WELL DEFINED TERTIARY "FUNDAMENTAL PARTICLE BY PRODUCT COMPOUNDS" (FP - BP). PRODUCE NEO GENESIS OF THE NEW DIFFERENT KIND ELECTRONS, NEUTRONS, PROTONS CONSTRUCTIONS, WHICH UNDER DEGENERATIVE A. S. I. FP. Mol. CIC. CAN REVERSE TO ORIGINAL CONSTRUCTIONS AS WELL.

IN EACH TEST THE CREATED NEW PARTICLE BY PRODUCTS SUCH AS RELEASED FREE FUNDAMENTAL PARTICLES INTO AIR OR SPACE, OR PRODUCED NEW PARTICLE COMPOUND BY PRODUCTS WHICH CONSTRUCT NEW ELECTRONS NUCLEONS PARTICLE COMPOUND CONSTRUCTIONS, ALSO ARE WELL DEFINED, CHEMICAL FORMULARY SPECIFIC, A. S. I. FP. Mol. CIC. SPECIFIC, AND EACH GIVEN PARTICLE BY PRODUCTS HAVE OWN CONSTRUCTION SPECIFIC MOLECULAR WEIGHTS, GRAVITON FORCES, AND POSSESS CHEMICAL FORMULARY SPECIFIC ENERGY CONTENTS.
 DEPENDING TO DIFFERENT TESTS ONE MAY PRODUCE BIOFRIENDLY SONIC OR LIGHT FUNDAMENTAL PARTICLES, IN CONTRAST THE BIOHOSTILE TESTS MAY CAUSE PRODUCTIONS OF

THE BIO LETHAL FREE FUNDAMENTAL PARTICLES SUCH AS X-PARTICLES, GAMMA OR LASER PARTICLES, ETC.

IN EACH TEST, A GIVEN EXOGENOUS PARTICLES WHICH IS FROM GIVEN SPECIFIC SUBCLASS THERMAL FUNDAMENTAL PARTICLE. VIA INTERNAL ATOM PARTICLE CIRCULATION SYSTEM (PCS). TRAVEL AND ENTER INTO SUBSYSTEM UNITS CHEMICAL LAB.S OF INTERNAL ELECTRONS AND NUCLEONS, AND COMBINE WITH INTERNAL ELECTRON NUCLEON EXISTING DIFFERENT KIND INDIGENOUS PARTICLE COMPOUNDS. UNDER A. S. I. FP. Mol. CIC.

A. S. I. FP. Mol. CIC. Between Lesser Weight, Lesser Energy, Lesser Graviton Force Particle Molecular Compounds

INTERNAL ELECTRON A. S. I. FP. Mol. CIC.

IN A. S. I. FP. Mol. CIC. ALWAYS, ONE NORMAL MOLECULE UNIT FROM A GIVEN FUNDAMENTAL PARTICLE MOLECULE WHICH POSSESSES LESSER PARTICLE WEIGHT (LESSER FP. MASS), LESSER GRAVITON FORCE, AND LESSER ENERGY CONTENT GIVEN FUNDAMENTAL PARTICLE COMPOUND COMBINES WITH ONE ANOTHER COMPATIBLE ONE NORMAL MOLECULAR WEIGHT LESSER GRAVITON FORCE, LESSER ENERGY CONTENT, AOTHER FUNDAMENTAL PARTICLE COMPOUND UNDER A. S. I. FP. Mol. CIC, AND THE COMBINATION PRODUCE TIRTIARY ONE NORMAL MOLECULAR WEIGHT LESSER ENERGY, LESSER GRAVITON FORCE, LESSER WEIGHT (LESS FP – MASS) TIRTIARY ANOTHER FUNDAMENTAL PARTICLES COMPOUNDS, WHICH IN THE ALL INTERACTIONS, THEIR PRDUCED PARTICLE COMPOUNDS ARE NORMAL MOLECULAR WEIGHT PARTICLE

COMPOUNDS. IN ORDER THEY CAN COMPLETE A GIVEN REGENERATIVE A. S. I. FP. Mol. CIC.

IN THESE TESTS MOSTLY THE LESSER THERMAL ENERGY CONTENT, LESSER GRAVITON FORCE AND LESSER PARTICLE WEIGHT EXOGENOUS THERMAL FUNDAMENTAL PARTICLES, MOSTLY COMBINE WITH INTERNAL ELECTRON INDIGENOUS WEAK ENERGY CONTENT, AND LESSER MOLECULAR WEIGHT EXISTING DIFFERENT INTERNAL SUBSYSTEM UNIT'S PARTICLE MOLECULAR SONIC PARTICLE COMPOUNDS. OR COMBINE WITH LESSER WEIGHT AND LESSER ENERGY CONTENT LIGHT FUNDAMENTAL PARTICLES COMPOUNDS. UNDER A. S. I. FP. Mol. CIC. INSIDE RECIPIENT ELECTRONS, THE CHEMICAL COMBINATIONS RESULTS THERMAL PARTICLE COMPOUND CONSTRUCTIONS.

THIS IS PROCESS OF THERMAL PARTICLE COMPOUND NEO GENESIS. AS WELL IT IS NEO ELECTRON GENESIS THROUGH STORING NEW THERMAL PARTICLE COMPOUNDS CONSTRUCTIONS.

AT ABOVE A. S. I. FP. Mol. CIC. THE EXOGENOUS THERMAL PARTICLES COMBINE WITH INTERNAL ELECTRON SONIC PARTICLE COMPOUNDS IN FIRST TEST AND PRODUCE NEW MOLECULAR COMPOUND CONSTRUCTIONS OF THERMAL PARTICLE COMPOUND, AND CAUSE RELEASE OF FREE SONIC PARTICLES TO OUT OF ATOM SPACE. IN NEXT TEST THE THERMAL PARTICLE COMBINE WITH INTERNAL ELECTRON LIGHT PARTICLE COMPOUND, AND CAUSE RELEASE OF LIGHT PARTICLES ASLO PRODUCE ANOTHER THERMAL PARTICLE COMPOUND MOLECULE INSIDE ELECTRONS.

INCOMING THERMAL EXOGENOUS PARTICLES REPLACE SONIC AND LIGHT FUNDAMENTAL PARTICLES IN NEW MOLECULAR COMPOUND CONSTRUCTIONS OF ELECTRONS. THE RELEASED

FREE SONIC FUNDAMENTAL PARTICLES AND RELEASED FREE LIGHT FUNDAMENTAL PARTICLES FROM ONE TEST TO NEXT, ONE BY ONE SCAPE INTENAL ATOM TO AIR AS FREE LIGHT AND SONIC FUNDAMENTAL PARTICLES, TRAVEL AIRBORNE. THESE RELEASED FREE PARTICLES ARE DETECTABLE IN TEST FIELD, CAN BE DETECTED, MEASURED.

T – FP + S. FP. – COMP. ------------→ T. FP. – COMP. + S – FP

←----------------

T - FP + Y. FP. – COMP. -----------→ T. FP. – COMP. + Y - FP

←----------------

The A. S. I. FP. Mol. ClC. between the Weak FP – Molecules

Combine exogenous external Atom weak light Weight (less weight) Thermal Fundamental Particles (T- FP) with Weak Indigenous Internal Electron Sonic (S-FP) or Light (Y-FP) Fundamental Particle Compounds of a given Metal Atoms under A. S. I. FP. Mol. ClC. will produce and release weak free S –FP, or weak free Y- FP, in the chemical combination results as the interaction Particle by Product (PBP). the produced weak light or sonic fundamental particles (PBP) will Scape into air and can be detected, sensed and disappear floating in air from lab.

Conclusion: The Exogenous Weak Thermal Fundamental Particle Molecules (that means which the combining FP – Molecules have lesser Particle Weights, lesser Graviton Force, and lesser Particle Energy content, when the EX. – T - FP enter into internal Electrons Subsystem Units chemical Lab. Combine with compatible weak internal Electron Y- FP – Compounds, or combine with Weak Internal Electron S- FP - compounds. The ongoing A. S. I. FP. Mol. CIC. cause releases of free S- FP, or free Y- FP into air, plus internal Electron Thermal –FP – Compounds as new PBP, which are reversible Chemical interactions.

A. S. I. FP. Mol. CIC. Between

Intermediate Particle Weight, Intermediate Energy, Intermediate Graviton Force Fundamental Particle Molecules

INTERNAL ELECTRON, NUCLEON A. S. I. FP. Mol. CIC.

==

THE INTERMEDIATE THERMAL ENERGY CONTENT, AND INTERMEDIATE QUANTUM QUANTITY PARTICLE WEIGHT EXOGENOUS THERMAL FUNDAMENTAL PARTICLES, TRANSMITTED INTO INTERNAL ATOM PCS.

INSIDE ATOM EXOGENOUS PARTICLES THROUGH INTERNAL ATOM PARTICLE CIRCULATION SYSTEMS (PCS) TRANSIT INTO PROTON, NEUTRON'S SUBSYSTEM UNIT CHEMICAL LAB.S, COMBINE WITH INTERNAL PROTON, NEUTRON COMPATIBLE INDIGENOUS INTERMEDIATE MOLECULAR

WEIGHT PARTICLE COMPOUNDS, (FOR EXAMPLE THE EXOGENOUS THERMAL PARTICLES COMBINE WITH A GIVEN INTERMEDIATE ENERGY CONTENT ELECTRIC PARTICLE COMPOUNDS). UNDER THE REGENERATIVE A. S. I. FP. Mol. CIC. AND PRODUCE THERMAL PARTICLE COMPOUND, PLUS FREE ELECTRIC PARTICLE.

THE A. S. I. FP. Mol. CIC. RESULT COMBINATION OF ABOVE TWO DIFFERENT PARTICLE MOLECULES WITH EACH OTHER, PRODUCE ANOTHER WELL DEFINED INTERMEDIATE WEIGHT AND THERMAL ENERGY INTERNAL NUCLEON THERMAL PARTICLE COMPOUND, ALSO RELEASE OF ELECTRIC FUNDAMENTAL PARTICLES AS FREE ELECTRIC CURRENT, TRAVEL INTO OUT OF ATOM. WHICH ARE DETECTABLE, AND CAN BE MEASURED AS RESULTS OF THE ABOVE A. S. I. FP. Mol. CIC.

IN ALL OF A. S. I. FP. Mol. CIC., ALWAYS ONE NORMAL MOLECULE UNIT FROM AN EXOGENOUS INTERMEDIATE WEIGHT FUNDAMENTAL PARTICLE COMBINE WITH ONE NORMAL MOLECULE UNIT, INTERNAL PROTON, NEUTRON, INTERMEDIATE ENERGY AND INTERMEDIATE WEIGHT PARTICLE COMPOUND, SUCH AS GIVEN ELECTRIC PARTICLE COMPOUND COMBINATION WITH THERMAL PARTICLES, UNDER A. S. I. FP. Mol. CIC. AND THE COMBINATION RESULT IS A PRODUCTION OF TERTIARY THERMAL PARTICLE COMPOUND WITH FREE ELECTRIC PARTICLES.

T – FP + E. FP. – COMP. ----------------→ E – FP + T. FP. – COMP.

The A. S. I. FP. Mol. CIC. between Standard Mediocre Industrial level FP - Molecules (S.M.I.L-FP)

Combine an Exogenous External Atom given S. M. I. L. – T- FP (Thermal – FP), which have mediocre T- FP weight, Mediocre Graviton Force, Mediocre Thermal Energy under A. S. I. FP. Mol. CIC. combine with the same previous Test Conditions Metal's Internal Electron – Nucleon S. M. I. L. Electric – FP – Compound Molecules, or other kind S. M. I. L. – FP – Compounds. these Chemical Combinations between these Molecules will produce the following Particle - By - Products (PBP) of freed S. M. I. L. – E. FP (mediocre weight Electric Fundamental particles), plus S. M. I. L. – T. FP – Compound Molecules inside the Recipient Metal's Electrons and Nucleons. And free released Electric Fundamental Particles as well as other similar mediocre released fundamental particle can be detected and recorded.

Conclusion: The Exogenous S. M. I. L. Thermal Fundamental Particle Molecules combine under A. S. I. FP. Mol. CIC. with S. M. I. L. internal Electrons *Nucleon's Electric or similar other types Fundamental particle compounds, and produce S. M. I. L. – T - FP. –Compounds inside recipient Atoms Internal Electrons Nucleons. Additionally, the freed Electric Fundamental Particles from A. S. I. FP. Mol. CIC. can be detected and measured as free Electric Fundamental Particles and Currents. Results of the Ex. S. M. I. L. - T- FP – Molecules Combinations with same weight FP and same level Graviton Force and Energy levels internal Atom Molecular Electric Fundamental Particle Compound molecules all A. S. I. FP. Mol. CIC. mostly takes place between the same fundamental Particle Weight and same level Graviton Force and Energy*

Molecules, produce the same level similar Weight and Graviton Molecules.

A. S. I. FP. Mol. CIC. between Heavy Weight, Higher Energy Content, Higher Graviton Force Fundamental Particles

INTERNAL PROTON, NEUTRON A. S. I. FP. Mol. CIC

==

IN THESE TWO TESTS, TWO DIFFERENT CLASS HIGHER ENERGY CONTENT, HIGHER WEIGHT, EXOGENOUS THERMAL FUNDAMENTAL PARTICLES, THROUGH INTERNAL PERIPHERON - NUCLEUS PCS, TRAVEL INTO DIFFERENT INTERNAL PROTON, NEUTRON SUBSYSTEM UNIT CHEMICAL LAB.S. UNDER A. S. I. FP. Mol. CIC. COMBINE RESPECTFULLY WITH TWO DIFFERENT INTERNAL PROTON - NEUTRON HIGHER ENERGY AND HEAVY WEIGHT COMPATIBLE X - PARTICLE AND GAMMA –PARTICLE MOLECULE- COMPOUNDS. TWO DIFFERENT TESTS PRODUCE TWO DIFFERENT KIND THERMAL FUNDAMENTAL PARTICLE COMPOUND CONSTRUCTIONS, AND RELEASED X – PARTICLES, AND GAMMA PARTICLES FROM THEIR PREVIOUS COMBINED

FORMS, INTO FREE FORMS PARTICLE, SCAPE INTERNAL ATOM SPACE TO OUTSIDE.

THE HEAVY WEIGHT, HIGH THERMAL ENERGY EXOGENOUS THERMAL PARTICLE COMPOUND CONSTRUCTIONS BUILD RECIPIENT ATOMS NUCLEUS, BY USING THE NEW HIGH ENERGY THERMAL PARTICLES, THERMAL PARTICLES REPLACE PRE-EXISTING HIGH ENERGY X- PARTICLES, AND GAMMA PARTICLES, RESPECTFULLY IN TWO DIFFERENT TESTS.

THIS IS PHENOMENON OF NEO PROTON GENESIS, AND NEO NEUTRON GENESIS. (OR NEO GENESIS OF NEW PROTONS, NEUTRONS AND ATOMS). THROUGH USING HIGHER ENERGY AND WEIGHT THERMAL PARTICLE COMPOUND CONSTRUCTIONS, AS WELL THIS IS CONSTRUCTION OF NEW THERMAL PARTICLES COMPOUND MOLECULES INSIDE PROTONS, AND NEUTRONS.

THESE A. S. I. FP. Mol. CIC. PRODUCE INTERNAL NEUTRON, PROTON HIGHER THERMAL ENERGY CONTENT, HIGHER WEIGHT, HIGHER GRAVITON FORCE NEW THERMAL PARTICLE COMPOUND CONSTRUCTIONS, ALSO THESE A. S. I. FP. Mol. CIC. RELEASE FREE FUNDAMENTAL PARTICLES SUCH AS X- PARTICLES, GAMMA PARTICLES, AND THESE FREED PARTICLES CAN BE DETECTED AND MEASURED.

T – FP + X. FP. –COMP. ----------------→ X – FP + T. FP. – COMP.

< ---------------

T – FP + GAMMA FP. – COMP. ----------→ GAMMA – FP + T. FP. – COMP.

< -------------------

THE A. S. I. FP. Mol. CIC between Vicious Lethally High Powerful Industrial Level FP. Molecules (V. L. H. P. I. L.– FP)

COMBINE EXOGENOUS EXTERNAL ATOM GIVEN V. L. H. P. I. L. – T. FP (THERMAL FUNDAMENTAL PARTICLES), WHICH ARE HEAVY WEIGHT T. – FP. AND HAVE HIGHER POWERFUL GRAVITON FORCE, VICIOUS HIGHER LETHAL ENERGY CONTENT THERMAL FUNDAMENTAL PARTICLE ENERGIES, UNDER A. S. I. FP. Mol. CIC. WITH THE SAME PREVIOUS TWO TESTS PATTERNS SIMILAR TEST – CONDITIONS, WITH METAL'S INTERNAL NEUTRONS PROTON'S V. L. H. P. I. L. GAMMA – FP – Mol. COMPOUNDS, OR X – FP – COMPOUNDS, OR WITH LASER – FP- Mol. - COMPOUNDS, ETC.

THESE CHEMICAL COMBINATIONS WILL PRODUCE INTERNAL NUCLEON'S V. L. H. P. I. L. –T. FP. – COMPOUNDS AS INTERACTION PARTICLE BY PRODUCTS (PBP), INSIDE THE RECIPIENT METAL ATOM'S NUCLEUS. ADDITIONALLY, THE PRODUCED AND RELEASED FREE GAMMA FUNDAMENTAL PARTICLES, OR FREED AND PRODUCED FREE X- FP, OR RELEASED FREE LASER FUNDAMENTAL PARTICLES FROM ABOVE MENTIONED REGENERATIVE OR DEGENERATIVE A. S. I. Mol. CIC. WILL SCAPE THE TEST AREA IN LAB. AS FREE FLOATING X- FP CURRENTS, OR GAMMA FUNDAMENTAL PARTICLE CURRENTS, OR LASER FUNDAMENTAL PARTICLE CURRENTS AS FREE PARTICLE CIRCULATION SYSTEMS AT AIR OR INTO SURROUNDING MEDIAS, AND CAN BE DETECTED AND MEASURED IN CONSEQUENCE.

CONCLUSION: THE V. L. H. P. I. L. FUNDAMENTAL PARTICLE MOLECULES A. S. I. FP. Mol. CIC. INSIDE THE A GIVEN ELEMENT'S

NUCLEUS INTWERNAL PROTONS NEUTRON'S SUBSYSTEM UNITS CHEMICAL LAB. WILL PRODUCE V. L. H. P. I. L. – FUNDAMENTAL PARTICLE COMPOUNDS AS WELL AS FREE VICIOUS X- PARTICLES, GAMMA PARTICCLES, LASER PARTICLES AS FINAL TEST'S INTERACTION'S PARTICLE MOLECULE -BY –PRODUCTS (PBP).

THE EX. – PCS & IND. – PCS FOR DOING: A. S. I. FP. Mol. CIC.

THE FUNDAMENTAL PARTICLE CLOUD MOLECULE CIRCULATION SYSTEMS (PCS) BETWEEN EXTERNAL ATOM EXOGENOUS THERMAL FUNDAMENTAL PARTICLE MOLECULES CIRCULATION SYSTEMS (EX. – PCS), WHEN CONNECTING TO INTERNAL ATOM INDIGENOUS PARTICLE CIRCULATION SYSTEMS OF INTERNAL ELECTRON NUCLEON PARTICLE CIRCULATION SYSTEMS CONSTRUCTIONS (IND-PCS).

THE EXOGENOUS THERMAL FUNDAMENTAL PARTICLES MOLECULES (EX – T. FP) TRAVEL EITHER AIRBORN OR THROUGH PERIPHERAL ATOM SPCES (PAS) EX. PCS CURRENTS AND FROM PERIPHERON UNDER ATTRACTIVE GRAVITON FORCES OF THE ELECTRONS AND NUCLEONS GET ATTRACTION INTO INTERNAL ATOM'S INDIGENOUS PARTICLE CIRCULATION SYSTEMS (IND-PCS), AND ENTER INTO INTERNAL ELECTRON, NEUTRON, AND PROTON SUBSYSTEM UNIT'S CHEMICAL LAB.S, AND UNDER REGENERATIVE A. S. I. FP. Mol. CIC. COMBINE WITH OTHER PRE-EXISTING DIFFERENT INTERNAL ELECTRON NUCLEON PARTICLE COMPOUNDS.

THE INTERNAL ELECTRONS, NUCLEON PARTICLE COMPOUNDS EXAMPLES ARE PARTICLE COMPOUNDS SUCH AS: SONIC PARTICLE MOLECULAR COMPOUNDS, LIGHT FUNDAMENTAL

PARTICLE COMPOUNDS, ELECTRIC PARTICLE COMPOUNDS, THERMAL FUNDAMENTAL PARTICLE COMPOUNDS, X- PARTICLE MOLECULAR COMPOUNDS, GAMMA PARTICLE COMPOUNDS, AND ETC.

UNDER REGENERATIVE A. S. I. FP. Mol. CIC. THE CHEMICAL COMBINATIONS OF THE INCOMING EXOGENOUS THERMAL FUNDAMENTAL PARTICLE WITH DIFFERENT VARIABLE PARTICLE WEIGHTS, DIFFERENT QUANTITIES THERMAL ENERGY CONTENTS, DIFFERENT QUANTUM QUANTITY GRAVITON ENERGY FORCES, ETC. IN EACH GIVEN TEST FOR EACH GIVEN THERMAL PARTICLE, WHEN COMBINING WITH COMPATIBLE ENERGY CONTENT DIFFERENT INTERNAL ELECTRON, NUCLEON DIFFERENT SONIC, ELECTRIC, LIGHT, GAMMA, X PARTICLE COMPOUNDS, AND PRODUCING NEW STRUCTURE THERMAL PARTICLE COMPOUNDS, AND RELEASING FREED GAMMA PARTICLES, X- PARTICLES, LASER FUNDAMENTAL PARTICLES, ELECTRIC PARTICLES, SONIC FUNDAMENTAL PARTICLES, LIGHT PARTICLES ETC. INTO OUTSIDE LAB. OR SPACE AS PARTICLE BY PRODUCTS OF THE A. S. I. FP. Mol. CIC.

WHEN THE LESSER THERMAL ENERGY, LESSER WEIGHT EXOGENOUS THERMAL PARTICLES COMBINE WITH INTERNAL ELECTRON LESSER ENERGY CONTENT AND LESSER WEIGHT COMPATIBLE SONIC PARTICLE COMPOUNDS, UNDER A. S. I. FP. Mol. CIC. PRODUCE NEW INTERNAL ELECTRON THERMAL PARTICLE COMPOUNDS, AND CONSTRUCT NEW CONSTRUCTION ELECTRONS, AT THE SAME TIME RELEASES SONIC PARTICLES AIRBORNE FREE AND CAN BE DETECTED, IN TEST FIELD.

WHEN THE LESSER THERMAL ENERGY, LESSER WEIGHT EXOGENOUS INCOMING THERMAL PARTICLES COMBINE WITH LESSER ENERGY AND LESSER WEIGHT INTERNAL ELECTRON LIGHT PARTICLE COMPOUNDS, PRODUCE LESSER ENERGY THERMAL PARTICLE COMPOUND INSIDE ELECTRONS, AND RELEASES THE

FREED LIGHT FUNDAMENTAL PARTICLES FREE INTO SPACE, WHICH CAN BE DETECTED AND MEASURED WHEN A. S. I. FP. Mol. CIC. COMPLETED.

"" THE A. S. I. FP. Mol. CIC. BETWEEN LESSER WEIGHT PARTICLE COMPOUNDS ARE MOSTLY INTERNAL ELECTRON TYPES CHEMICAL INTERACTIONS.""

WHEN THE A. S. I. FP. Mol. CIC. TAKE PLACE BETWEEN INTERMEDIATE ENERGY CONTENT EXOGENOUS THERMAL PARTICLES, WHEN THEY ARE COMBINING WITH INTERNAL ELECTRON NUCLEON INTERMEDIATE WEIGHT AND INTERMEDIATE ELECTRIC ENERGY, AND INTERMEDIATE GRAVITON FORCE ELECTRIC PARTICLE COMPOUNDS INSIDE THE ELECTRON, NUCLEON SUBSYSTEM UNIT CHEMICAL LAB.S, THESE CHEMICAL INTERACTIONS PRODUCE INTERMEDIATE ENERGY CONTENT THERMAL PARTICLE COMPOUNDS INSIDE NEUTRONS, AND PROTONS.

"" MOSTLY THESE INTERACTIONS ARE NUCLEAR (INTERNAL PROTON, INTERNAL NEUTRON TYPES A. S. I. FP. Mol. CIC.""

THE FREED INTERMEDIATE WEIGHT, AND INTERMEDIATE ELECTRIC ENERGY CONTENT FREED ELECTRIC FUNDAMENTAL PARTICLES CURRENTS AND PARTICLES SCAPE INTERNAL ATOM SPACE, CAN BE DETECTED AND MEASURED.

WHEN THE HIGHER WEIGHT, HIGHER THERMAL ENERGY CONTENT, HIGHER GRAVITON FORCE EXOGENOUS INCOMING THERMAL FUNDAMENTAL PARTICLES INSIDE ATOM'S SUBSYSTEM UNIT CHEMICAL LAB.S COMBINE WITH INTERNAL PROTON, INTERNAL NEUTRON HIGHER ENERGY X- PARTICLE COMPOUNDS, OR HIGHER WEIGHT AND HIGHER GAMMA ENERGY CONTENT GAMMA PARTICLE COMPOUNDS, ETC. UNDER A. S. I. FP. Mol. CIC.

THESE CHEMICAL COMBINATION PRODUCE FREE GAMMA PARTICLES, X-PARTICLES FUNDAMENTAL PARTICLE CURRENTS

KNOW AS BEAMS PRESENTLY, DEPENDING WHAT ENERGY QUANTITY CONTENTS THERMAL PARTICLES, COMBINED WITH WHICH KIND DIFFERENT ENERGY CONTENT INTERNAL NUCLEON DIFFERENT GAMMA OR X- PARTICLE COMPOUNDS. AND CAUSE PRODUCTION OF NEW HIGHER THERMAL ENERGY THERMAL PARTICLE COMPOUNDS AS WELL. THESE A. S. I. FP. Mol. CIC. ARE INTERNAL NUCLEUS (INTERNAL NEUTRON AND INTERNAL PROTON) CHEMICAL INTERACTION CYCTES AND CHAINS OF THE DIFFERENT KINDS.

NANO NEUROLOGY OF LIVING THINGS

==
========

How PIS Units Control & Operate Total Body Electron, Neutron, Proton Population's Functions in Plants and Animals

EFFERENT AND AFFERENT PARICLE CLOUD CIRCULATION SYSTEMS

Ascending and Descending Bidirectional Particle Cloud Circulation Systems

========================

====

(Vertical or Longitudinal PCS)

FUNCTIONS OF O – PIS, S – PIS, CNS – PIS
=============================

The Plants, Animals Body Organs and Body Systems each one possesses independent Organ Specific, System Specific Particle Intelligence System Centers (PIS), which are in charges of body Operations of the physical, chemical biological body function. Each individual Organ and System Particle intelligence system centers (PIS) are capable to emit and receive Order Particle Clouds from self, as well as the PIS Nano Units are capable to receive and analyze Particle Clouds from other centers, and operate all different body organs A. S. I. FP. Mol. CIC.

The emitted particle clouds are capable to travel through Total Body Particle Circulation Systems (PCS), also the Particle Clouds and individual fundamental particle currents also are capable to travel and transit inside Electrons, Nucleons, Atoms, through internal Atom Particle Circulation System Routes (A-PCS). Each given atom, Electron, Nucleon have Nano Unit specific PIS, which are in charge of the different total Body Nano unit function operations, these different PIS also connect to each other through general body PCS operations.

Each given individual internal Cell inter Cytoplasmic or internal Cell Nuclear Organs and Systems also have independent PIS. The internal Cell Particle Circulation Systems (PCS) connect all of these different internal nucleus and internal Cytoplasmic PIS to each other.

IN LIVING THINGS, THE BODY ORGAN'S PARTICLE INTELLIGENCE SYSTEM CENTERS NANO UNITS (O – PIS) OPERATE, AND CONTROLS ORGAN'S ELECTRONS, PROTONS, NEUTRONS ATOMS, BIOMOLECULES, AND CELLS (PHYSICAL, CHEMICAL, BIOLOGICAL) – FUNCTIONS. THE O – PIS ARE CAPABLE TO EMIT THEIR RESPONE, AND ORDER PARTICLE CLOUDS, AS WELL AS THE O – PIS ARE CAPABLE TO RECEIVE DIFFERENT TYPES INCOMING PARTICLE CLOUDS, FROM OTHER BODY ORGANS AND SYSTEMS, AS WELL AS FROM ENVIRONMENTAL OUTSIDE ORIGIN PARTICLE CLOUDS.

IN THE LIVING THINGS, THE S- PIS (BODY SYSTEM'S PARTICLE INTELLIGNECE SYSTEMS CENTERS NANO UNITS) ARE IN CHARGE OF OPERATION OF BIOLOGICAL, CHEMICAL, PHYSICAL, MECHANICAL, ELECTRICAL, THERMAL AND MANY OTHER DIFFERENT OTHER TYPES FUNCTIONS OF DIFFERENT BODY STSTEM ELECTRONS, NEUTRONS, PROTONS, ATOMS, BIOMOLECULES, CELLS FUNCTION ORPERATIONS. THE S – PIS ARE CAPABLE TO EXCHANGE CLOUDS.

THE DIFFERENT CNS – PIS NANO UNIT CENTERS (CNS – PIS), ALSO ARE IN CHARGE OF OPERATION OF BODY ORGANS, AND BODY SYSTEM'S TOTAL ELECTRONS, NUCLEON, ATOM POPULATIONS PHYSICAL, CHEMICAL, BIOLOGICAL FUNCTION OPERATIONS. THE CNS – PIS NANO UNITS ARE DOMINANT NANO UNIT CENTERS, THROUGH EMISSION OF

ORDER PARTICLE CLOUDS CONTROL TOTAL O – PIS, AND S – PIS, CONSEQUENTLY ORDER AND CONTROL TO TOTAL BODY ELECTRON, NUCLEON, ATOM POPULATION FUNCTIONS. THROUGH EMITTING AND RESPONDING BY CLOUD OPERATIONS. THE O- PIS, AND S – PIS BOTH ARE RECESSIVE TOWARD CNS – PIS ORDER CLOUDS.

UNDER NANO UNIT HIERARCHY SYSTEMS, THE CNS ELECTRONS, PROTONS, NEUTRONS, AND ATOMS ARE DOMINANT NANO UNITS AGAINST ALL OTHER REMAINING BODY SYSTEMS AND BODY ORGANS ATOMS, ELECTRONS, NUCLEONS POPULATIONS.

PARTICLE CIRCULATION SYSTEMS (PCS)
DOMINANT AND RECESSIVE PIS NANO UNITS

THE TOTAL BODY ATOMS, ELECTRONS, NUCLEON PARTICLE INTELLIGENCE SYSTEM CENTERS (PIS) THROUGH EMITTING AND RECEIVING S. Y. T. E. – FP – I.I. – P. Cl. VIA PARTICLE CIRCULATION SYSTEMS CONNECT EXISTING INFROMATION AND IMAGE PARTICLE CLOUDS TOTAL BODY CELL, TOTAL BODY ATOM POPULATIONS TO EACH OTHER, UNDER HIERARCHY CLOUD CIRCULATIONS AND ORDER OPERATIONS, BY PCS CONNECTIONS.

THE BODY ORGAN PARTICLE INTELLIGENCE SYSTEM CENTERS (O – PIS), THE BODY SYSTEMS PARTICLE

INTELLIGENCE SYSTEM CENTERS (S – PIS), AND CNS – PIS, UNDER HIERARCHY SYSTEM CENTERS EMIT PARTICLE CLOUDS (P. Cl.) FROM THEIR INTER ELECTRON, NEUTRON, PROTON, ATOM CONSTRUCTIONS, UNDER A. S. I. FP. Mol. CIC. AND THROUGH PARTICLE CLOUD CIRCULATION SYSTEMS THE OPERATION OF THE ENTIRE BODY ELECTRON AND NUCLEON, AND ATOM AND CELL OPERATIONS CONTROLLED AND ACHIEVED UNDER NANO UNIT HEIRARCHY SYSTEM.

THE EMITTED ORDER PARTICLE CLOUDS FROM ANY PIS, ENTER INTO GENERAL BODY ORGANS AND BODY SYSTEM'S PARTICLE CLOUD CIRCULATION SYSTEMS (PCS), THE ORDER PARTICLE CLOUDS ENTER INTO RECIPIENT O – PIS, S – PIS, CNS – PIS. AND UNDER A. S. I. FP. Mol. CIC. INTERACT WITH RECIPIENT NANO UNITS PIS -ANALYSIS. THE RECIPIENT CNS – PIS, O - PIS AND S – PIS RESPOND, ACT, FUNCTION, ACCORDINGLY UNDER HIERARCHY SYSTEMS CLOUD ORDERS. ADDITIONALLY, EMIT AND PRODUCE RESPONSE PARTICLE CLOUDS INTO GENERAL PCS.

DOMINANT CNS - PIS NANO UNITS, CONTROL THE OPERATIONS OF RECESSIVE BODY ORGANS AND BODY SYSTEMS ELECTRONS, NEUTRONS, PROTONS AND ATOMS FUNCTIONS THROUGH THE EMITTED ORDER PARTICLE CLOUDS INSTRUCTIONS, AND OPERATIONS.

HEIRARCHY OPERATION BETWEEN BODY ELECTRONS NUCLEONS O – PIS, S – PIS, CNS – PIS OPERATE TOTAL BODY FUNCTIONS ATOM- CELL HIERARCHY SYSTEM OPERATIONS

THE PARTICLE INTELLIGENCE SYSTEMS OF THE BODY ORGANS (O – PIS), THE PARTICLE INTELLIGENCE SYSTEMS OF THE BODY SYSTEMS (S – PIS), AND THE BRAIN (CNS – PIS) OPERATE AND CONTROL ALL BODY ORGANS AND BODY STSEM'S ELECTRON, NEUTRON, PROTON, ATOM' TOTAL POPULATIONS, AND TOTAL BODY CELL POPULATION'S BIOLOGICAL PHYSIOLOGICAL AND CHEMICAL FUNCTIONS.

THE PERIPHERAL S- PIS, AND O- PIS SEARCH, DISCOVER, PRODUCE PARTICLE CLOUD REPORT AND TRANSMIT PARTICLE CLOUD INFORMATION AND IMAGE REPORTS (Y. S. T. E. – FP – I.I. – P. Cl.) FROM PHERIPHERAL BODY ORGANS AND SYSTEM'S ELECTRONS, NUCLEONS, ATOMS, CELL'S PHYSICAL, CHEMICAL, BIOLOGICAL FUNCTION OPERATION REPORTS TO BRAIN (CNS – PIS) ATOMS, CELLS, PROTONS, NEUTRONS, AND ELECTRONS OBSERVATIONS AND DECISIONS.

THE PARTICLE CLOUDS FROM PERIPHERAL ORGANS TO CNS-PIS TRAVEL THROUGH PARTICLE CLOUD QUANTUM

CIRCULATION SYSTEMS BACK AND FORTH BETWEEN CNS – PIS, S – PIS, AND O- PIS, THROUGH BI -DIRECTIONAL FLOWING PARTICLE CLOUD CURRENTS OF EFFERENT AND AFFERENT PARTICLE CLOUD CURRENTS.

THE O – PIS, AND S – PIS CONTINUOSLY REPORT NORMAL, ABNORMAL, PHYSIOLOGICAL, PATHOLOGICAL ONGOING EVENTS PARTICLE CLOUDS INFORMATIONS AND IMAGE CLOUDS TO CNS – PIS ELECTRONS NUCLEONS, BEFORE DIFFERENT CNS – PIS CENTERS, BACK AND FORTH, UNDER THE HIERARCHY ATOMS, ELECTRONS, NUCLEONS AND CELL'S OPERATION SYSTEMS.

ALSO, S – PIS, O – PIS REPORT EXTERNAL WORLD ENVIRONMENTAL INFORMATIONS AND IMAGES PARTICLE CLOUD TO CNS – PIS DIFFERENT CENTERS ELECTRONS, NEUTRONS, PROTONS, ATOMS, AND CELLS, FOR THEIR A. S. I. FP. Mol. CIC. ACTIONS. THE CNS – PIS AND BRAINS ELECTRON, NEUTRON, PROTON, ATOM, AND CELL POPULATIONS ARE IN NEED OF THESE INFORMATION AND IMAGES PARTICLE CLOUDS, IN ORDER TO DECIDE CORRECTLY AND ISSUE CORRECT PARTICLE CLOUD ORDERS, UNDER A. S. I. FP. Mol. CIC.

THE BODY SYSTEMS TOTAL ELECTRON, NUCLEON POPULATION'S PHYSICAL, CHEMICAL, BIOLOGICAL FUNCTIONS CONTROLLED UNDER ORDERS OF BODY SYSTEMS PARTICLE INTELLIGENCE SYSTEM CENTERS (S – PIS). THROUGH STRICT ORDERS AND COOPERATION OF INCOMING CLOUD ORDERS FROM SUPERIOR HIERARCHY CNS – PIS NANO UNIT CENTERS CLOUD ORDERS.

ENTIRE BODY SYSTEMS, SUCH AS G.I.S., G.U.S., CARDIO PULMONARY SYSTEMS, CNS, ENDOCRINE SYSTEMS, CUTANEOUS AND MUSCULOSKELETAL SYSTEMS, ETC. ALL HAVE SYSTEM SPECIFIC NANO UNIT PARTICLE INTELLIGENCE SYSTEM CENTERS (S – PIS). ABOVE FACTS ARE TRUE IN REGARD TO ALL OTHER LIVING THING SPECIES S – PIS, O-PIS, CNS - PIS AS WELL.

CNS – PIS ELECTRONS, NUCLEON, ATOM AND CELL POPULATION ARE DOMINANT BODY INTELLIGENCE SYSTEM CENTERS. CNS - PIS CONTROL ENTIRE BODY ORGANS AND BODY SYSTEMS ELECTRONS, NUCLEONS, ATOMS AND CELLS UNDER HIERARCHY NANO UNIT OPERATION SYSTEMS. THE BRAIN ELECTRONS, NUCLEONS, ATOMS AND CELLS ARE COMMANDER NANO UNIT POPULATIONS. THE REMAINING TOTAL BODY ELECTRONS, NUCLEONS, NANO UNITS OF BODY MOSTLY AS RECESSIVE ORGAN'S CELLS, ATOMS, NANO UNITS MUST WORK UNDER THE CNS – PIS PARTICLE CLOUD ORDER SYSTEMS.

UNDER HIERARCHY NANO UNIT SYSTEMS OPERATION, THE RECESSIVE O - PIS WHICH HAVE LESSER NANO UNIT ENERGY, AND GRAVITON ELECTRONS, NEUTRONS, PROTONS AND ATOMS MUST FUNCTION UNDER CNS – PIS CLOUD ORDERS DOMINANCES.

THE CNS ELECTRONS NEUTRONS PROTONS AND ATOMS UNDER A. S. I. FP. Mol. CIC. PRODUCE AND EMIT ORDER Y. S. E. T. - FP. - I.I. - P. Cl. AND OTHER PARTICLE CLOUD ORDERS, THESE EMITTED ORDER PARTICLE CLOUDS ENTER INTO GENERAL PCS BIDIRECTIONAL PARTICLE CURRENT FLOWS, AND TRANSIT INTO PERIPHERAL ORGANS, AND SYSTEMS

DIFFERENT PIS CENTERS. THE RECESSIVE – PIS HAVE COOPERATE AND COORDINATE FUNCTIONS WITH CNS – PIS ORDER CLOUDS INSTRUCTIONS.

THE DOMINANT CNS ELECTRONS NUCLEONS, UNDER CNS CELLS, ATOM'S DOMINANCE HIERARCHY BEHAVIORS EMIT DIFFERENT ORDER PARTICLE CLOUDS INTO BODY PCS. THE CNS – PIS NANO UNITS UNDER THESE ORDER PARTICLE CLOUDS, OPERATE AND CONTROL ENTIRE PLANTS, LIVING THINGS, AND ANIMALS TOTAL BODY POPULATION ATOMS, ELECTRONS, PROTONS, NEUTRON PHYSICAL, CHEMICAL, BIOLOGICAL FUNCTIONS.

THE PERIPHERAL ORGANS AND THE PERIPHERAL BODY SYSTEMS ATOMS, ELECTRONS, NUCLEONS, AND PARTICLE INTELLIGENCE SYSTEM CENTERS (O – PIS, S – PIS) ALSO HAVE INDEPENDENT AUTONOMOUS DECISION MAKING CAPABILITIES AND PARTICLE CLOUD EMISSION AND CLOUD COMMUNICATION ABILITIES.

THE INCOMING DESCENDING INFORMATION IMAGE PARTICLE CLOUDS FROM BRAIN ATOMS ORIGINS, ENTER INTO RECIPIENT O – PIS, AND S – PIS INTERNAL ELECTRON NUCLEON SUBSYSTEM UNIT, UNDER A. S. I. FP. Mol. CIC. INTERACT WITH PARTICLE COMPOUNDS CONSTRUCTIONS OF INSIDE O – PIS, S – PIS NANO UNITS, ELECTRONS AND NUCLEONS, RELEASE RESPONSE PARTICLE CLOUDS, ORDERS AS WELL, THROUGH ASCENDING PCS, THESE ORGAN CLOUDS RETURN BACK TO CNS – PIS, FOR THEIR INTERACTIONS.

THE PERIPHERAL O- PIS, AND S – PIS EMIT THEIR RESPONSE PARTICLE CLOUD, ABOUT ALL EXISITNG PERIPHERAL

EXISTED INFROMATIONS AND IMAGES IN PARTICLE CLOUD FORMS, AND PERIPHERAL O- PIS, S – PIS ISSUED PARTICLE CLOUD DECISIONS, AS WELL AS ALL KIND OTHER TYPES Y. S. E. T. – FP- I.I. – P. Cl. RESPONSE- CLOUDS, FINALLY SENT BACK AND TRANSITTED TO CNS – PIS. THROUGH REVERSE DIRECTION PCS FLOWS, BACK TO HIGHER CNS - PIS.

THE RESPONSE CLOUDS ENTER INTO REVERSE PCS CURRENTS, AND TRAVEL BACK THROUGH REVERSE PCS ROUTES BACK TO THE O- PIS, S- PIS, FINALLY BACK TO THE CNS – PIS CENTERS FOR THEIR FINAL DECISION AND FINAL COMMANDING ORDERS.

Above are Communication of total body Electron, Neutron, Proton, Atom and Cells Population with each other, and Operation of living things Life under hierarchy Nano Unit systems.

THE BODY ORGANS AND BODY SYSTEMS, HAVE SYSTEM SPECIFIC PARTICLE CIRCULATION SYSTEM (S - PCS), AND

ORGAN SPECIFIC PARTICLE CIRCULATION SYSTEMS (O – PCS).

THE TOTAL BODY ELECTRON NUCLEON POPULATIONS CONNECT TO EACH OTHER, THROUGH EFFERENT AND AFFERENT PARTICLE CLOUD CURRENT ROUTES (PCS). DIFFERENT OUT GOING AND INCOMING PARTICLE CLOUD CURRENTS ORDERS FROM CNS TO PERIPHERAL ATOMS. OR FROM PERIPHERAL ELECTRONS NUCLEONS TOWARD CNS ATOM CENTERS. ALL COORDINATE AND EXCHANGE THEIR PARTICLE CLOUD INFORMATION AND IMAGES, WHICH CONNECT TO EACH OTHER THROUGH FUNDAMENTAL PARTICLE CLOUD CIRCULATION SYSTEMS.

THE ABOVE PHENOMENON IS, TOTAL BODY P. Cl. – PCS, TOTAL BODY PIS - P. Cl. OPERATION. WHICH THEY OPERATE TOTAL BODY ELECTRON, NEUTRON, PROTON, ATOM, CELL, TISSUE, ORGAN, SYSTEM BIOLOGICAL, CHEMICAL, PHYSICAL FUNCTIONS AND TOTAL LIVING THING BODY OPERATIONS ALL TOGETHER WITH FULL COORDINATION AND COOPRERATION OF EACH OTHER. THROUGH FUNDAMENTAL PARTICLE PRECISION ACCURACIES.

Fundamental Particle Diseases

THE BODY ORGANS FUNCTION AND THE BODY SYSTEMS FUNCTION

UNDER "ORDER – CLOUDS OF THEIR PIS"

CARDIAC – ORGAN WORKS UNDER CARDIAC - O -PIS ORDERS AND BRAIN'S AUTONOMOUS - CNS – PIS ORDERS, PLUS CARDIAC - O - PCS

CARDIAC ORGAN'S WORKS CONTROLLED AND DONE UNDER ORDER CLOUDS OF: CARDIAC O– PIS, AND CNS – PIS

THE CARDIAC AURICLE LOCATED "CARDIAC ORGAN PARTICLE INTELLIGENCE SYSTEM CENTER (C- O – PIS), TOGETHER WITH BRAIN'S CNS - AUTONOMOUS - PIS CENTERS (CNS – A – PIS) JOINTLY OPERATE FUNCTION OF TOTAL CARDIA- CELL POPULATIONS, AS WELL AS THESE TWO PIS- CENTERS OPERATE AND CONTROLS FUNCTIONS OF TOTAL CARDIAC ATOM'S POPULATIONS, TOTAL CARDIAC ELECTRONS AND NUCLEONS POPULATIONS, AND CARDIAC DIFFERENT TISSUE- OPERATION AND TASKS, BY CLOUD ORDERS, WHICH CIRCULATED UNDER BI- DIRECTIONAL ORDER CLOUDS CURRENT AT CARDIAC PARTICLE CIRCULATION SYSTEMS (C- PCS).

ADDITIONALLY, UNDER SAME OPERATION PATTERNS THE DIFFERENT BODY ORGAN'S PARTICLE INTELLIGENCE SYSTEM CENTERS (O – PIS), TOGETHER WITH VOLUNTARY AND INVOLUNTARY BRAIN CENTERS HIERARCHY PARTICLE CLOUD OPERATION ORDER CLOUD SYSTEMS (CNS – PIS – CLOUD OPERATIONS) CONTROL ENTIRE TOTAL BODY CELL'S POPULATIONS. AS WELL AS THESE DIFFERENT PIS OPERATE THE TOTAL BODY ATOMS, ELECTRONS, PROTONS,

NEUTRONS TOTAL BODY POPULATIONS UNDER THE PARTICLE CLOUD OPERATIONS, THE PARTICLE CLOUD CURRENTS TRANSIT AND TRAVEL THROUGH GENERAL BODY PARTICLE CIRCULATIONS SYSTEMS (PCS), WHICH ARE BI DIRECTIONAL ASCENDING OR DESCENDING PARTICLE CIRCULATION SYSTEMS AS WELL AS TRANSVERCE PARTICLE CIRCULATION SYSTEMS CURRENTS.

UNDER C- PIS AND CNS- PIS HIERARCHY PARTICLE CLOUD ORDERS, WHICH TRANSIT THROUGH CARDIAC PCS THE HEART ORGANS TOTAL ATOMS, ELECTRONS, NEUTRONS, PROTONS AND CELLS AND TISSUES FUNCTION TWENTYFOUR HOURS PER DAY FOR OVER 50 – 120 YEARS WITHOUT STOPS, ACCURATELY AND PERFECTLY WITHOUT ANY MISTAKES, THROUGH FUNDAMENTAL PARTICLE PRECISION ACCURIES, AND PERFORM ALL CHEMICAL, BIOLOGICAL, PHYSIOLOGICAL, MECHANICAL, THERMO-ELECTRIC, ELECTRO-DYNAMIC, ETC. CARDIAC ORGAN FUNCTIONS ALL TIMES. PROVIDE NEEDED PARTICLE MOLECULES, AND NEEDED ELEMENTAL MOLECULES TO TOTAL BODY CELLS, AND ATOMS CHEMICAL LAB.S, WHICH WITHOUT CARDIAC BIOLOGICAL PUMPING TASK LIFE STOP TO CONTINUE IN MATTERS OF MINUTES.

ANY MALFUNCTION IN CNS – PIS FUNCTION, AND ANY CARDIAC – PIS CENTER'S IRREGULARITIES, OR ANY DISORDER IN PCS FUNCTION, OR DEFECT IN PARTICLE CLOUD ORDERS, AND PARTICLE CLOUD CIRCULATION SYSTEM ALL CAN DISRUPT CARDIAC FUNCTIONS, PRODUCE DYSRHYTHMIA, ARREST, FUNCTION DISORDER, TISSUE DAMAGES, AND CAN THREATEN THE LIFE.

UNDER MOLECULAR EVOLUTION ORDERS, THE ATOMS PIS, AND CELLS PIS RESEARCH DISCOVERED AND CONSTRUCTED THIS ELECTRO MECHANICAL PUMPING BIO DEVICE, ACCORDING MOLECULAR EVOLUTION ORDERS AND PATHS. IN ORDER TO PROVIDE NEEDED MOLECULES FOR BIOLOGICAL MACHINE'S TOTAL BODY POPULATION NANO UNITS, AND MICRO UNITS.

THE OUTSIDE ENVIRONMENTAL EVENTS AND INCIDENTS THROUGH PRODUCED ABNORMAL PARTICLE CLOUD CIRCULATION SYSTEMS CAN DISRUPT CARDIAC PCS, AND CAN DISRUPT CARDIAC PIS FUNCTIONS.

ALSO THE DISEASES OF OTHER BODY ORGANS, THROUGH TRANSMITTED ABNORMAL PARTICLE CLOUDS AND ABNORMAL PCS CONNECTIONS, ALL AFFECT CARDIAC PARTICLE CLOUD CIRCUALTION OPERATIONS, MALFUNCTIONS OF THE CARDIAC PIS, AND PRODUCE DISORDERS IN CARDIAC PARTICLE CLOUD CIRCULATION SYSTEM (O – PCS), IN DISTURBANCES IN CARDIAC - PIS, ALSO AT CNS – PIS, AND CNS – PCS, AND ALL CAN PRODUCE CARDIAC FUNCTION DISORDERS AND CARDIAC DAMAGES.

UNDER REVERSE PARTICLE CLOUD MISTAKE REPORTS FROM PERIPHERY IN OPPOSITE DIRECTION, REVERSE PCS, THE CARDIAC PARTICLE INTELLIGENCE SYSTEM CENTER (O – PIS) ALSO CAN MALFUNCTIONS ALSO CAN AFFECT BRAINS CNS – PIS FUNCTIONS. RESULT SECONDAY CHAIN CARDIAC – CNS OPERATION MALFUNCTIONS INTERACTIONS AND DISORDERS.

 THE AUTONOMOUS CNS – PIS CENTER CONTROL CONTINUOUSLY THE CARDIAC – PIS CENTERS FUNCTION.

THE VOLUNTARY CNS – PIS ALSO CAN CONTROL AND AFFECT CARDIAC PARTICLE CIRCULATION SYSTEMS, CARDIAC - PIS.

THERE ARE NUMEROUS OTHER ENVIRONMENTAL EVENTS, THAT OCCURRENCES OF THOSE INCIDENTS THROUGH PARTICLE CIRCULATION SYSTEMS CONNECTIONS CAN PRODUCE CARDIAC O – PIS MALFUNCTION, CNS – PIS DYSFUNCTION, CAN PRODUCE O - PCS DYSRHYTHMIA, ARREST, RHYTHM IRREGULARITIES, MALFUNCTIONS AND DISORDERS, AND CAN CAUSE INSTANT O – PCS AND CARDIAC O – PIS ARREST, AND SECONDARILY CARDIAC ORGAN INJURY, MI. ETC.

Particle Cloud produced Diseases of OTHER ORGANS

IN ALL OTHER BODY ORGANS, THE PARTICLE INTELLIGENCE SYSTEM CENTER (O – PIS) OF THE ORGAN AND BRAIN JOINTLY CONTROL AND OPERATE BODY ORGANS TOTAL ELECTRONS, NUCLEONS, ATOMS, BIOMOLECULES, CELLS FUNCTIONS IN DIFFERENT PHYSICAL, CHEMICAL, BIOLOGICAL FUNCTION OPERATIONS OF THE ENTIRE LIVING THINGS BODY ORGAN AND SYSTEMS EXACTLY SIMILAR AS EXPLAINED FOR HEART AND CARDIOVASCULAR SYSTEM.

MALFUNCTIONS OF PARTICLE CLOUDS, OR PARTICLE INTELLIGENCE SYSTEMS (PIS), OR DISORDERS OF THE PARTICLE CIRCULATION SYSTEMS (PCS) IN THE SAME BODY ORGAN AND SYSTEM OR EVEN IN ANOTHER BODY ORGANS AND SYSTEMS ALL INFLUENCE THE ELECTRONS, NUCLEONS,

ATOMS, CELLS, TISSUES FUNCTIONS, AND CAN PRODUCE IMMENSE NUMBER DIFFERENT KIND DISEASES IN ALL BODY SYSTEMS. AND CAN PRODUCE DIFFERENT ORGANS RELATED SIGNS AND SYMPTOMS SUCH AS; IN RESPIRATORY SYSTEMS THESE DISORDERS MAY PRODUCE CONDITIONS SIMILAR TO ASMTHMA, CHEST PAINS, DYSPNEA, COUGH AND MANY OTHER TYPES OF SYNDROMES, WHICH MANY OF THEM WRONGFULLY MIS NAMED UNDER IDIOPATHIC OR AUTOIMMUNE DISEASES, ETC.

ORGAN FUNCTIONS UNDER HEIRARCHY CNS- PIS AND PERIPHERAL ORGAN'S PARTICLE INTELLIGENCE SYSTEMS (O - PIS) ACHIEVE BUILDING, Syntheses of Proteins, Enzymes, Hormones, Biomolecules, Cells, under A. S. I. Mol. CIC. AT DIFFERENT BODY SYSTEMS AS: GIS, GUS, ETC.

THE REGENERATIVE AND DEGENERATIVE AUTONOMOUS SEQUENTIAL CHEMICAL INTERACTION CHAINS AND CYCLES (A. S. I. Mol. CIC.) FOR CONSTRUCTION OF CELLS, BODY TISSUES NEED SPECIFIC WELL DEFINED CHEMICAL FORMULARY MOLECULES, AND BIOMOLECULES. IN ORDER TO USE THOSE MOLECULES AND CONSTRUCT INTERNAL CELL'S ORGANS AND SYSTEMS.

THERE ARE BODY ORGANS AND SYSTEMS SUCH AS: GIS, LIVER, PANCREAS, SPLEEN, BONE MARROW, GUS, LUNGS, ETC. ARE THE ORGANS UNDER THEIR PARTICLE INTELLIGENCE SYSTEM PARTICLE CLOUD ORDERS, WHO MAKE DECISIONS, UNDER PARTICLE INTELLIGENCE SYSTEMS ADVISE AND DECISIONS PRODUCE PARTICLE CLOUDS FOR THE FUNCTIONS WHICH NEEDED TO BE DONE BY OTHER ORGANS OR INSIDE DIFFERENT CHEMICAL LAB.S OR THROUGH MECHANICAL FORCES, AND BIOLOGICAL TASKS.

UNDER THE PIS DOMINANT ORDERS AND PRODUCED PARTICLE CLOUD ORDERS THE DIFFERENT BODY ORGANS AND SYSTEMS PERFORM THE BIOLOGICAL, CHEMICAL, PHYSICAL TASKS, AND ACHIEVE THE NEEDED WORKS FOR BIOLOGICAL MACHINES. UNDER PIS PARTICLE CLOUD ORDERS LIVING THINGS EVEN CAN COLLECT NECESSITIES EVEN FROM OUTSIDE ENVIRONMENT AND DELIVER THE NEEDED FOOD MOLECULES OR VITAL GASES SUCH OXYGEN TO RELATED BODY SYSTEMS TO BE USED FOR DIFFERENT ACTS.

THE GASTROINTESTINAL (GIS) CELLS AND NANO UNIT'S PARTICLE INTELLIGENCE SYSTEMS CENTERS (O – PIS) WITH COORDINATION OF ORDERS FROM CNS O – PIS, AND O- PCS THROUGH A. S. I. Mol. CIC. PRODUCE THE NEEDED SPECIFIC MOLECULES FOR CONSTRUCTION OF CELLS. AND THROUGH CIRCULATION SYSTEMS PROVIDE TO NEEDED MATERIALS FOR BODY ORGANS CHEMICAL LAB.S WHO ARE IN NEEDS OF THE MATERIALS.

UNDER O – PIS ORDERS THE GIS, LIVER, SPLEEN, PANCREAS, BONE MARROW, THROUGH AUTONOMOUS SEQUENTIAL INTER MOLECULAR CHEMICAL INTERACTION CHAINS AND CYCLES (A. S. I. Mol. CIC.) CONSTRUCT ATOMS, ELECTRONS, NUCLEONS, CELLS, INTERNAL CELL'S ORGANS AND SYSTEMS CONSTRUCTIONS SUCH AS: DUPLICATIONS OF CHROMOSOMES, MITOCHONDRIA, BIOMOLECULAR DUPLICATION SYSTEMS, CELL WALL, BUILDING CELL'S INTELLIGENCE SYSTEM CENTERS, CELL'S IMMUNITY AND DEFENSIVE SYSTEMS, ETC. CONSEQUENTLY, CONSTRUCT CELLS, TISSUES, BODY ORGANS, BODY SYSTEMS, LIVING THINGS, ETC. AND PRODUCE DUPLICATE PLANT AND ANIMAL LIVING THINGS, UNDER PARTICLE INTELLIGENCE SYSTEM CENTERS PARTICLE CLOUD ORDERS. THROUGH USING OF COLLECTED MATERIALS FROM THE OUTSIDE ENVIRONMENT. THIS IS PHENOMENON OF CELL GENESIS, CELL SYSTEM GENESIS, CELL ORGAN GENESIS UNDER CONTROLS OF O- PIS, CNS – PIS THROUGH A. S. I. Mol. CIC.

PHENOMENON OF CELL GENESIS

INTRODUCTION TO INTERNAL CELL ORGANS AND SYSTEMS

MOST CELLS HAVE BEEN CONSTRUCTED FROM TWO SECTIONS EACH HAS INDEPENDENT PIS CENTER, THE DOMINANT HEAVY CELL NUCLEUS SEGMENT WITH MORE GRAVITON FORCE AND ENERGY, UNDER PIS CLOUD ORDERS TAKE CENTRAL POSITION LIKE ANY OTHER DOMINANT

HEAVY MASS, AND FROM THE CENTRAL NUCLEAR POSITION UNDER STRONG ATTRACTIVE GRAVITON FORCES OF CELL NUCLEI ATTRACT FROM PERIPHERY UNDER GRAVITON FORCE ATTRACT THE WEAK, LESS MASS AND WEAK GRAVTON FORCE CYTOPLASMIC ORGANS AND SYSTEMS MASSES. AND UNDER PARTICLE CLOUD ORDERS OF PIS PRODUCE BICOMPARTMENT CELL CONSTRUCTION. THIS IS PHENOMENON OF BICOMPARTMENT CELL GENESIS.

THE NUCLEAR CENTRAL SEGMENT OF CELL, AND THE CYTOPLASMIC PERIPHERAL SEGMENT OF THE CELL, BOTH HAVE BEEN CONSTRUCTED FROM DIFFERENT INTERNAL CELL ORGANS AND INTERNAL CELL SYSTEMS. SUCH AS CELL'S BIDIRECTIONAL EFFERENT AND AFFERENT CIRCULATION SYSTEMS, OR CELL'S TRANSPORATION SYSTEMS (EITHER BRINGS IN NEEDED MOLECULES, OR TAKING OUT NON NEEDED MOLECULES), CELL'S IMMUNITY AND DEFENSIVE SYSTEMS, CELL'S INTELLIGENCE SYSTEM CENTERS, CELL'S DUPLICATION AND DIVISION CHEMICAL LAB.S, CELL'S ENERGY PRODUCER AND PROVIDER SYSTEMS, CELL'S PERIPHERAL PROTECTIVE AND DEFENSIVE ORGANS, (CELL MEMBRANE), CHROMOSOMES, MITOCHONDRIA (CHEMICAL LAB.S, AND ENERGY PRODUCER SYSTEMS), ETC. WHICH ALL SYSTEMS FUNCTION UNDER O – PIS, CNS – PIS, AND O – PCS PARTICLE CLOUD ORDERS].

THE GIS, GUS, LUNGS, LIVER, SPLEEN, BONE MARROW, ETC. ARE THE ORGANS AND SYSTEMS WHICH UNDER REGENERATIVE AND DEGENERATIVE AUTONOMOUS SEQUENTIAL CHEMICAL INTERACTION CHAINS AND CYCLES

PRODUCE AND PROVIDE WELL DEFINED NEEDED MOLECULES AND BIOMOLECULES FOR CELL GENESIS, TISSUE GENESIS, BODY ORGAN AND SYSTEM CONSTRUCTION, AND GENESIS OF DIFFERENT SPECIES LIVING THINGS, UNDER THE O – PIS, CNS – PIS, O - PCS.

THE O – PIS, CNS – PIS, O- PCS ARE IN CHARGE OF OPERATION OF AUTONOMOUS SEQUENTIAL CHEMICAL INTERACTION CYCLES AND CHAINS (A. S. I. Mol. CIC.) OF GIS. GUS. LIVER, SPLEEN, PANCREAS, BONE MARROW, LUNGS, ETC. IN ORDER TO CONSTRUCT NEEDED MOLECULES AND BIOMOLECULES WHICH THROUGH USE OF THOSE TO CONSTRUCT CELL ORGANS, CELL SYSTEMS, CELLS, TISSUES AND BODY ORGANS CONSTRUCTIONS, ETC.

THE O – PIS, CNS – PIS, O- PCS ARE THE COMMANDING OPERATORS TO PRODUCE, AND PROVIDE NEEDED MOLECULES, BIOMOLECULES, CHROMOSOMES, ETC. FOR CELL CONSTRUCTIONS, TISSUE GENESIS, AND LIVING THING MOLECULAR CONSTRUTIONS, THROUGH A. S. I. Mol. CIC.

FOR EXAMPLE, GASTROINTESTINAL SYSTEM, PANCREAS, LIVER, SPLEEN, BONE MARROW, ETC. UNDER DEGENERATIVE AND REGENERATIVE AUTONOMOUS SEQUENTIAL CHEMICAL INTERACTION CHAINS AND CYCLES (A. S. I. Mol. CIC.), EITHER BREAK DOWN THE LARGE FOOD MOLECULES INTO SMALL ONES, OR IN OPPOSITE REVERSE DIRECTIONS A. S. I. Mol. CIC. CONSTRUCT LARGE MOLECULES FROM COMBINATION OF SEVERAL SMALLER ONES, IN ORDER TO MAKE READY NEWLY CONSTRUCTED MOLECULES, WHICH TO USE THESE NEW CONSTRUCTED MOLECULES AND BIOMOLECULES FOR BUILDING INTERNAL

CELL ORGANS, CELL SYSTEMS, BIOMOLECULES, CHROMOSOMES, MITOCHONDRIA, AND OTHER INTERNAL CELL SYSTEMS, AND INTERNAL CELL ORGANS, INSIDE DIFFERENT CELL'S CHEMICAL LAB.S, UNDER O – PIS, AND CNS – PIS PARTICLE CLOUD ORDERS, THROUGH A. S. I Mol. CIC. UNDER EXACT SAME ORDERS OF MOLECULAR EVOLUTION PATHS.

THESE ARE PHENOMENON OF CELL ORGAN'S GENESIS, AND CELL SYSTEM GENESIS, AND TISSUE GENESIS PROCESS UNDER PAST MOLECULAR EVOLUTION ORDERS. BY A. S. I. Mol. CIC. ORDERS, REPEATS EXACTLY THE SAME AT PRESENT ERA.

Bio Molecule Genesis, the Gene Genesis
The Cell and Tissue Genesis Phenomenon
IN MOLECULAR EVOLUTION PATHWAYS

THE PARTICLE INTELLIGENCE SYSTEM CENTER OF DOMINANT ORGANS (O – PIS, CNS - PIS) DECIDE, PRODUCE AND EMIT ORGAN SPECIFIC PARTICLE CLOUDS (P. Cl.), THE PRODUCED P. Cl. THROUGH PARTICLE CIRCULATION SYSTEM (PCS) TRAVEL INTO RECIPIENT CELL'S INTERNAL ELECTRON - NUCLEON - PIS, AS WELL AS TRAVEL INTO RECIPIENT CELL'S O – PIS CENTERS, AND RECIPIENT CELL SYSTEM'S – PIS CENTERS.

THE RECIPIENT PARTICLE INTELLIGENCE SYSTEM CENTERS (RECIPIENT - PIS) HAS TO FUNCTION ACCORDING PARTICLE CLOUDS ORDER INCLUDING RECIPIENT ELECTRONS -PIS, THE NUCLEONS AND ATOM'S - PIS FUNCTIONS ACCORDING THE CLOUD ORDERS, AS WELL AS THE RECIPIENT CELLS AND TISSUE PIS OF DIFFERENT BODY ORGANS AND SYSTEMS, ALL INTERNAL CELL ORGANS AND INTERNAL CELL SYSTEMS ALSO HAS TO FUNCTION PHYSICALLY, CHEMICALLY AND BIOLOGICALLY UNDER CLOUD ORDERS. INN GENERAL DOMINANT PIS UNDER HIERARCHY CLOUD SYSTEMS CONTROL THE FUNCTIONS OF THE DIFFERENT RECIPIENT CHEMICAL LAB.S. CELLS, TISSUE, ATOMS, ETC.

THE RECIPIENT PIS – CENTERS ACCORDING ORDERS OF PARTICLE CLOUD EXECUTE MOLECULAR, BIOMOLECULAR PRODUCTIONS EXACTLY THE SAME AS HAS BEEN ORDERED BY THE CLOUDS, AND PRODUCE NEEDED, MOLECULES, GENES, BIOMOLECULES, HORMONES, ENZYMES, CELL ORGANS, CELL SYSTEMS, CELLS, AND TISSUES. ACCORDING THE INSTRUCTIONS OF PARTICLE CLOUDS INSIDE CHEMICAL LAB.S, THROUGH A. S. I. Mol. CIC. UNDER THE PATHS AND ORDERS OF THE MOLECULAR EVOLUTION.

THESE ARE PHENOMENON OF THE HORMONE GENESIS, ENZYME GENESIS, GENE GENESIS, CHROMOSOME GENESIS, ELECTRONS AND NUCLEON GENESIS, UNDER MOLECULAR HIERARCHY SYSTEMS OF THE DOMINANT ELECTRONS, NUCLEONS, PARTICLES, ATOMS, CELLS, CELL SYSTEMS AND CELL ORGANS PARTICLE INTELLIGENCE SYSTEM CENTERS PARTICLE CLOUD ORDERS.

Repeating Molecular Evolution

LIVE MOLECULE GENESIS UNDER A. S. I. Mol. CIC.

NEO CELL ORGAN GENSIS, CELL GENESIS, TISSUE GENESIS, LIVING THING CONSTRUCTION, AND NEO ATOM GENESIS.

UNDER DOMINANT MOLECULES AND DOMINANT CLOUD ORDERS, UNDER CNS – PIS CLOUDS, AND O – PIS CLOUD ORDERS.

CNS – PIS, O – PIS EMIT PARTICLE CLOUDS (P. Cl.) AND TRANSIT THESE P. Cl., INTO PARTICLE CLOUD CIRCULATION SYSTEMS (PCS). THE PARTICLE CLOUD RECIPIENT ORGAN'S INTERNAL CELL CHEMICAL LAB.S, THROUGH ORDERS OF THESE EXOGENOUS PARTICLE CLOUDS (EX. - P. Cl.), FUNCTION ACCORDINGLY, THE RECIPIENT ORGAN'S A. S. I. Mol. CIC, & A. S. I. FP. Mol. CIC., UNDER CLOUD ORDERS CONSTRUCT NEW BIOMOLECULES, INTERNAL CELL ORGANS, AND INTERNAL CELL SYSTEMS, AS WELL INTERNAL ELECTRON NUCLEON NEW PARTICLE COMPOUND CONSTRUCTIONS. PRODUCE STRUCTURES SUCH AS HORMONES, ENZYMES, RNA, DNA, CHROMOSOMES, GENES, INTERNAL CELL ORGANS, AND INTERNAL CELL SYSTEMS, SUCH AS MITOCHONDRIA, CELL MEMBRANE, CELL'S

TRANSPORTATION SYSTEMS, GENE DUPLICATION CHEMICAL LAB.S, AND MANY OTHER LIVE AND NON LIVE NEW MOLECULAR CONSTRUCTIONS.

ALSO UNDER EXOGENOUS PARTICLE CLOUD ORDERS, INSIDE ELECTRON, NUCLEON CHEMICAL LAB.S THROUGH A. S. I. FP. Mol. CIC. PRODUCE NEW PARTICLE COMPOUND CONSTRUCTION, NEO ELECTRON GENESIS, NEW NUCLEON AND ATOM CONSTRUCTIONS. ACCORDING O- PIS, CNS – PIS, ISSUED ORDER PARTICLE CLOUD, THROUGH HIERARCHY PARTICLE CLOUD MOLECULAR OPERATIONS. INTERNAL CELL DOMINANT MOLECULE GENESIS, ENZYME GENESIS ALSO ARE ANOTHER EXAMPLES.

THESE DOMINANT MOLECULES SUCH AS HORMONES, ETC. HAS BEEN CONSTRUCTED FOR USE IN ANOTHER FAR AWAY ORGANS, THESE DOMINANT MOLECULES EXCRETE INTO BODY CIRCULATION SYSTEMS, TRAVEL INTO TARGET ORGANS AND SYSTEMS INTERNAL CELL AND INTERNAL ATOM CHEMICAL LAB.S.

THE RECIPIENT CHEMICAL LAB.S. UNDER DOMINANT MOLECULES AND BIOMOLECULES EMITTED PARTICLE CLOUD ORDERS, THROUGH WELL DEFINED FIX A. S. I. Mol. CIC. INSIDE RECIPIENT INTERNAL CELL, AND INTERNAL ELECTRON NUCLEON CHEMICAL LAB.S CONSTRUCT NEEDED LIVE AND NON LIVE MOLECULAR PARTICLE COMPOUND CONSTRUCTION, NEO ELECTRON GENESIS, AND NEO NUCELON GENESIS, CONSTRUCTIONS. ADDITIONALLY, UNDER DOMINANT MOLECULAR HORMONES AND OTHER DOMINANT BIOMOLECULAR STRUCTURE PRODUCED AND ORDERED PARTICLE CLOUD, THE RECIPIENT CELLS AT

INTERNAL CELL CHEMICAL LAB.S PRODUCE AND CONSTRUCT NEEDED CELL ORGAN CONSTRUCTIONS, AND INTERNAL CELL SYSTEM CONSTRUCTIONS, OR ORDINARY SIMPLE NEEDED MOLECULES SUCH AS ENZYMES, BIOMOLECULES, CHROMOSOMES, GENES, RNA, DNA STRUCTURES, CELL ORGANS, AND CELL SYSTEMS, WHICH HAD BEEN ORDERED AND CONTROLLED BY CNS – PIS, AND O – PIS, ADDITIONALLY UNDER DOMINANT MOLECULAR ORDERS ACCORDINGLY AS WELL.

THE INTERNAL CELL ORGANS AND SYSTEMS, THE INTERNAL CELL ORGAN'S CHEMICAL LAB.S, AS WELL AS THE INTERNAL CELL SYSTEMS CHEMICAL LAB.S, A. S. I. Mol. CIC. ALL FUNCTION UNDER DOMINANT MOLECULAR AND BIOMOLECULAR ORDER CLOUD SYSTEMS AND INSTRUCTIONS.

THE DOMINANT MOLECULES SUCH AS HORMONES ALSO OPERATE THE A. S. I. Mol. CIC. INSIDE INTERNAL CELL CHEMICAL LAB.S, AS WELL AS THE DOMINANT PARTICLE COMPOUND MOLECULAR CONSTRUCTIONS ALSO CONTROL INTERNAL ELECTRON, INTERNAL NUCLEON CHEMICAL LAB.S OPERATIONS AND CONTROL A. S. I. FP. Mol. CIC. OPERATIONS INSIDE THE ATOMS AS WELL. AND PRODUCE NEEDED LIVE AND NON LIVE MOLECULAR CONSTRUCTIONS SUCH AS ENZYMES, DNA, RNA, CHROMOSOME, GENE, INTERNAL CELL ORGANS AND SYSTEMS, ALSO INSIDE ATOMS PRODUCE ELECTRONS AND NUCEONS NEW MOLECULAR COMPOUND CONSTRUCTIONS. ACCORDING CNS – PIS, AND O- PIS CLOUD ORDERS. THESE ARE PHENOMENONS OF HORMONE GENESIS, ENZYME GENESIS, RNA- NDA NEO GENESIS, GENE GENESIS, CELL AND TISSUE

GENESIS PROCESS. ALSO AT INTERNAL ATOM CONSTRUCT NEW ELECTRON NUCLEON CONSTRUCTIONS (ATOM NEO GENESIS), WHICH ALL OF THESE NEW CONSTRUCTIONS ARE UNDER MOLECULAR EVOLUTION ORDERS AND PATHS.

THE ALL PRESENTLY ONGOING AUTONOMOUS SEQUENTIAL CHEMICAL INTERACTIONS UNDER MONINANT MOLECULES, DOMINANT BIOMOLECULES, CELLS, ETC. WHICH PRESENTLY OCCUR AND CONSTRUCT EXISTING LIVE AND NON LIVE EXISTING SUBJECTS ACROSS THE UNIVERSE AT PRESENT TIME, ALL OF THESE NEO- GENESIS PHENOMENONS ALL ARE EQUAL TO THE PAST MOLECULAR EVOLUTION ORDERS. AND PATHS. THE ONGOING EXISTING THINGS GENESIS PHENOMENONS AL ARE AT FOOT STEPS OF THE PAST MOLECULAR EVOLUTION ORDERS AND PATHS.

"D. M. – PIS, D. - BM. – PIS"

"DOMINANT MOLECULES AND BIOMOLECULES- PIS"

Dominant Molecule's – PIS Relationships
with S – PIS, O – PIS, CNS – PIS, PCS, P. Cl.
under A. S. I. FP. Mol. CIC., A. S. I.- BM.- CIC.

THE ELECTRONS, NEUTRONS, PROTONS AND ATOMS OF DOMINANT BIO MOLECULES AND NON LIVE DOMINANT MOLECULES (FOR EXAMPLE HORMONES) HAVE BEEN CONSTRUCTED FROM SPECIFIC TYPES FUNDAMENTAL PARTICLE MOLECULES, PARTICLE COMPOUNDS, AND PARTICLE COMPOSITE CONSTRUCTIONS.

THE DIFFERENT DOMINANT MOLECULE'S ELECTRONS, NUCLEONS HAVE BEEN CONSTRUCTED FROM DIFFERENT PARTICLE CLASSES, SUCH AS ELECTRIC PARTICLE COMPOUNDS, THERMAL, LIGHT, SONIC FUNDAMENTAL PARTICLE COMPOUND CONSTRUCTIONS (ETC.).

ADDITIONALLY, THERE ARE MANY OTHER PRESENTLY KNOWN OR UNKNOWN OTHER FUNDAMENTAL PARTICLES COMPOUND CONSTRUCTIONS, WHICH HAVE BEEN USED IN CONSTRUCTIONS OF DOMINANT MOLECULES.

THE DOMINANT MOLECULES THROUGH PARTICLE CLOUD EMISSION FROM SELF CONTROL AND OPERATE DIFFERENT RECIPIENT BODY ORGAN'S – PIS FUNCTIONS, AND CELL'S PIS OPERATIONS. THE CELL'S ELECTRONS, NUCLEONS BIOLOGICAL, CHEMICAL, PHYSICAL FUNCTIONS ALL CONTROLLED BY DOMINANT MOLECULES ORDER PARTICLE CLOUD OPERATION SYSTEMS. THE RECIPIENT INTERNAL CELL ORGAN'S PIS, AND RECIPIENT INTERNAL CELL BIOMOLECULE'S PIS FUNCTIONS ALL CONTROLLED BY DOMINANT MOLECULES FUNDAMENTAL PARTICLE CLOUD ORDER.

DOMINANT MOLECULES AND BIOMOLECULE POSSESSES CLOSE FUNCTION OPERATION COOPERATION AND COORDINATION WITH PARTICLE CLOUD ORDERS OF O- PIS, S – PIS, CNS – PIS, O- PCS, S – PCS.

THE DOMINANT MOLECULES PARTICLE CLOUD ORDERS JOINTLY WORK WITH DIFFERENT BODY ORGAN'S PIS CHEMICAL, BIOLOGICAL, PHYSIOLOGICAL FUNCTIONS, AND A. S. I. Mol. CIC. OPERATIONS TOGETHER. THE DOMINANT MOLECULES AND BIOMOLECULES INTELLIGENCE SYSTEM CENTERS WITH COOPERATION AND COORDINATION OF CNS – PIS, O – PIS, PCS, ALTOGETHER OPERATE AND CONTROL DIFFERENT INTERNAL BODY ORGAN'S INTERNAL CELLS AND INTERNAL ELECTRON, INTERNAL NUCLEONS AND NANO UNITS PHYSICAL, CHEMICAL BIOLOGICAL FUNCTIONS,

THROUGH FUNDAMENTAL PARTICLE ACCURACY PRECISIONS.

DOMINANT MOLECULES AND BIOMOLECULES ARE DIFFERENT CONSTRUCTION NANO UNIT, MICRO UNIT CONSTRUCTIONS. THE DOMINANT MOLECULES INTERNAL ELECTRON NUCLEON STRUCTURES HAS BEEN CONSTRUCTED FROM DIFFERENT CHEMICAL FORMULARY FUNDAMENTAL PARTICLE COMPOUNDS. THE DOMINANT MOLECULES OPERATE BIOLOGICAL FUNCTIONS THROUGH JOINT CLOUD ORDERS WITH O – PIS, CNS – PIS, O- PCS. P. Cl. OPERATION CENTERS.

THE DOMINANT MOLECULES PARTILCE CLOUD ORDERS, AND CNS – PIS, O – PIS PARTICLE CLOUD ORDERES ALL JOINTLY OPERATE RECIPIENT ORGAN'S INTERNAL ELECTRON, NUCLEON PARTICLE COMPOUND CONSTRUCTIONS, ATOM NEO GENESIS, AND CELL NEO GENESIS IN RECIPIENT ORGANS. THROUGH A. S. I. FP. Mol. CIC. AND A. S. I. Mol. CIC.

THE PARTICLE CLOUD ORDER SYSTEMS OPERATE INTERNAL CELL ORGAN GENESIS, INTERNAL CELL SYSTEM GENESIS, INTERNAL CELL NUCLEUS GENESIS, AND INTERNAL CYTOPLASMIC MOLECULAR AND BIOMOLECULAR CONSTRUCTIONS. THROUGH PARTICLE PRECISION ACCURACIES, WITH EXPONENTIAL AUTONOMOUS SEQUENTIAL CHEMICAL INTERACTION CYCLES AND CHAIN SPEEDS, WITH NO MISTAKES.

THE INTERNAL CELL ORGAN GENESIS, AND INTERNAL CELL SYSTEM GENESIS SUCH AS: MITOCHODRIA CONSTRUCTIONS, GENE DUPLICATION SYSTEMS, INTERNAL

CELL CHEMICAL LAB. CONSTRUCTIONS, CHROMOSOME GENESIS, CONSTRUCTION OF INTERNAL CELL INTELLIGENCE CENTER SYSTEMS, AND INTERNAL CELL TRANSPORTATION SYSTEM CONSTRUCTIONS, GENESIS OF INTERNAL CELL IMMUNITY AND DEFENSIVE SYSTEMS, PERIPHERAL PROTECTIVE DEFENSIVE SYSTEM CONSTRUCTION FOR CELLS (CELL WALL GENESIS), AND ETC.

CONSTRUCTION OF THE ALL OF THESE DIFFERENT INTERNAL CELL ORGANS AND INTERNAL CELL DIFFERENT SYSTEMS ALL ARE UNDER CELL'S PIS, CNS – PIS, PCS CLOUD ORDER OPERATIONS, AND DOMINANT MOLECULAR PIS CONTROLS. ALL DIFFERENT CELL, BIO MOLECULE, MOLECULE CONSTRUCTION GENESIS PHENOMENONS, ALL ARE CONSTRUCTED AS IN EAXCT MOLECULAR COPIES OF MOLECULAR EVOLUTION ORDERS AND PATHS.

MOLECULR EVOLUTION = LIVING THING GENESIS

THE POST PLANETARY SYSTEM EXPLOSION, INCLUDING POST EXPLOSION OF GALAXIES, UNIVERSE, PRODUCED DUST. WHICH THE DUST IS FLOATING FUNDAMENTAL PARTICLES, NANO UNITS, MOLECULES, DIFFERENT SIZE MASSES, ROCKS, FLUIDS, PLASMA, ETC.

IN PLANET EARTH, THEREAFTER IN POST DUST INCIDENT ERA, THE MOLECULAR EVOLUTION TRIGGER STARTED ENORMOUS DIFFERENT KIND A. S. I. FP. Mol. CIC., AND A. S. I. Mol. CIC. UNDER ORDERS OF NANO UNIT'S PIS, PCS. OF PRODUCED FUNDAMENTAL PARTICLES, ELECTRONS, NUCLEONS, ELEMENTS, DIFFERENT MOLECULAR MASSES, UNDER PARTICLE INTELLIGENCE SYSTEM CENTERS PARTICLE CLOUD ORDERS OPERATIONS OF MASS UNITS, AS FOLLOWING:

(First phase Molecular Evolution)

Molecular Evolution in Fundamental Particles and Nano Units, (till Bio Molecule -Genesis)

DURING THIS EVOLUTION ERA, UNDER ORDERS OF EMITTED PARTICLE CLOUDS FROM NANO UNIT'S PARTICLE INTELLIGENCE SYSTEM CENTERS (PIS), THE DUST MOLECULES (FUNDAMENTAL PARTICLES, ELECTRONS, NEUTRONS, PROTONS, ATOMS, AND OTHER NANO UNITS), UNDER AUTONOMOUS SEQUENTIAL INTER NANO UNIT CHEMICAL INTERACTION CHAINS AND CYCLES, ALSO UNDER A. S. I. FP. Mol. CIC. COMBINED WITH EACH OTHER, AND PRODUCED LARGER CONSTRUCTION DIFFERENT NANO UNIT MOLECULAR COMPOUND SIZES, AS WELL AS LARGER MOLECULAR FUNDAMENTAL PARTICLE COMPOUNDS, AND

PRODUCED OTHER LARGER QUANTUM MOLECULAR STRUCTURES UP TO BIO MOLECULE LEVEL NEW MASS UNITS.

DURING THIS EVOLUTION ERA, THE PARTICLE INTELLIGENCE SYSTEM CENTERS (NANO - PIS) OF DIFFERENT NANO UNIT STRUCTURES {SUCH AS ELECTRONS, NEUTRONS, PROTONS, ELECTRON –NUCLEON'S SYSTEM UNITS (EXAMPLE QUARKS), ELECTRON NUCLEON SUBSYSTEM UNITS, ATOMS, ETC} WERE IN CHARGE OF ONGOING PHYSICAL, CHEMICAL, BIOLOGICAL FUNCTIONS, A. S. I. FP. Mol. CIC. AND AUTONOMOUS INTER NANO UNIT CHEMICAL INTERACTION CHAINS AND CYCLES, AND CONSTRUCTED NEW DIFFERENT SIZES OF OTHER QUANTUM SIZE MOLECULAR CONSTRUCTIONS, NEW NANO UNIT DISCOVERIES UP TO BIO MOLECULAR GENESIS LEVEL OF THE EVOLUTION.

Molecule Genesis take place under A. S. I. FP. Mol. CIC. as well as under A. S. I. NU. CIC. (Nano Unit = NU). The Molecule Genesis at Present Era is equal to Molecule Genesis during the Evolution Era.

MICRO UNIT (MIC. U.) EVOLUTION PHASE, OR SECOND ERA MOLECULAR EVOLUTION IN PLANET EARTH

EVOLUTION IN MICRO UNITS (Micro. U.)

EACH GIVEN CELL (ONE EXAMPLE FROM MICRO UNIT) HAVE BEEN CONSTRUCTED FROM NUMEROUS DIFFERENT INTERNAL CYTOPLASMIC AND INTERNAL NUCLEI DIFFERENT CELL ORGANS, CELL SYSTEMS SUCH AS: CELL'S PIS, CELL'S PCS, CELL'S MOLECULAR AND BIOMOLECULAR TRANSPORTATION SYSTEMS (BI-DIRECTIONAL EFFERENT AND AFFERENT MOLECULAR TRANSPORTATION SYSTEMS), CELL'S PARTICLE TRANSPORTATION SYSTEMS ARE BIDIRECTIONAL PARTICLE TRANSMISSION ROUTES, INTERNAL CYTOPLASMIC AND INTERNAL CELL NUCLEI INTELLIGENCE SYSTEM CENTERS, INTERNAL CELL CYTOPLASMIC AND INTERNAL CELL NUCLEI MOLECULAR AND BIOMOLECULAR DUPLICATION SYSTEMS, CELL'S IMMUNITY AND DEFENSIVE SYSTEMS, CELL'S ENERGY PRODUCTION LAB.S, CELL'S CHEMICAL LAB.S, AND MANY OTHER INTERNAL CELL ORGANS AND SYSTEMS WHICH ALL OF THESE ORGANS AND SYSTEMS INDIVIDUALLY POSSESSES SEPARATE INDEPENDENT PARTICLE INTELLIGENCE SYSTEMS CENTERS, PCS, AND FUNCTION UNDER CELL- PIS PARTICLE CLOUD ORDERS.

DURING MICRO UNIT ERA MOLECULAR EVOLUTION, THE ALL MICRO UNIT'S CONSTRUCTION DISCOVERIES WHICH IT INCLUDES THE DIFFERENT INTERNAL CELL SYSTEM DISCOVERIES, AND INTERNAL CELL ORGANS CONSTRUCTION DISCOVERIES TOOK PLACE, AND DISCOVERED AND CONSTRUCTED BY CELL'S AND BIOMOLECULES PARTICLE INTELLIGENCE SYSTEMS CENTERS ORDER PARTICLE CLOUD OPERATIONS.

THE DIFFERENT INTERNAL CELL ORGAN MICRO UNIT CONSTRUCTIONS DISCOVERED AND BUILD UNDER RESEARCH, FINDINGS, AND CONSTRUCTIONS BY CELL'S PARTICLE

INTELLIGENCE SYSTEM CENTERS DECISIONS AND CONSTRUCTIONS, AND ONE DISCOVERY AFTER THE OTHER, BUILT THROUGH A. S. I. Mic. U. CIC. UNDER PIS, THROUGH THE EVOLUTION ORDERS. TODAYS ONGOING MICRO UNIT GENESIS ALSO IS EQUAL TO ELVOLUTION MICRO UNIT GENESIS PROCESS. (EXCEPT WHEN THERE ARE NEW OTHER MICRO UNIT MUTATIONS TAKING PLACE).

ABOVE IS BRIEFLY MICRO UNIT'S EVOLUTION CONSTRUCTIONS UNDER PIS, PCS, THROUGH PARTICLE CLOUD ORDERS. THE PIS OF BIO MOLECULES, THE PIS OF THE CELLS, AND CELL ORGAN'S PIS, THE CELL SYSTEM'S PARTICLE INTELLIGENCE SYSTEM CENTERS (PIS), EMIT ORDER PARTICLE CLOUDS, AND THE EMITTED PARTICLE CLOUDS THROUGH INTERNAL CELL PARTICLE CIRCULATION SYSTEMS TRAVEL TO RECIPIENT INTER CELL ORGANS AND SYSTEMS.

ACCORDING PARTICLE CLOUD ORDERS, THE RECIPIENT CELL ORGANS AND SYSTEMS PRODUCE CELL ORGANS AND CELL SYSTEMS FUNCTIONS UNDER PIS - CLOUD ORDERS, ALSO RECIPIENT CELL ORGANS AND SYSTEMS CONSTRUCT CLOUD ORDERED BIOMOLECULES, RNA, DNA OR OTHER INTERNAL CELL MOLECULAR, BIOMOLECULAR CONSTRUCTIONS, GENE DUPLICATIONS, CELLS AND TISSUE GENESIS UNDER A. S. I. Mol. CIC. ACCORDING THE PARTICLE CLOUD ORDERS.

THE CELL'S PIS ADDITIONALLY ARE CONTROLLED BY OTHER SUPERIOR PIS CENTERS, AS WELL BY INFERIOR PIS ISSUED PARTICLE CLOUD REPORTS. CNS – PIS CENTERS IS DOMINANT PARTICLE CLOUD ISSUING CENTER WHICH CONTROLS AND WATCH TO THE FUNCTIONS OF ENTIRE BODY CELL POPULATIONS, CONTROLS ENTIRE BODY ATOM POPULATIONS FUNCTIONS, CNS – PIS CONTROLS AND REGULATES ENTIRE TOTAL BODY ELECTRONS, NEUTRONS, PROTONS, AND NANO UNITS POPULATION FUNCTIONS.

THE MICRO UNIT PATRICLE INTELLIGENCE SYSTEM CENTERS OPERATE NOT ONLY THE STABLE ONGOING ENTIRE BIOLOGICAL, PHYSICAL, CHEMICAL FUNCTIONS OF INTERNAL CELL ORGANS AND SYSTEMS. ASLO THEY ARE IN CONTROL OF ALL NEWLY CONSTRUCTED STABLE MUTATIONS, NEO MICRO UNIT GENESIS PHENOMENONS.

THE DIFFERENT INTERNAL CYCTOPLASMIC AND INTERNAL CELL NUCLEI PARTICLE INTELLIGENCE SYSTEM CENTERS, THROUGH PARTICLE CLOUD COMMUNICATION SYSTEMS INTERACT AND COMMUNICATE WITH EACH OTHER. THE DISCOVERED NEW CELLS AND TISSUE CONSTRUCTIONS, THE CORRECTION OF PREVIOUSLY CONSTRUCTED CELL MISHAPS AND TISSUE CONSTRUCTION MISTAKES ALL INTERACT BETWEEN THE DIFFERENT PIS CENTERS THROUGH PCS, AND PARTICLE CLOUD OPERATION SYSTEMS. THE PIS ACHIEVE AND CONTROL PRODUCTION OF ALL WELL DEFINED, RELATIVELY FIX CHEMICAL FORMULARY CONSTRUCTION MOLECULES, BIO MOLECULES, RNA, DNA, GENE, CHROMOSOME, AND OTHER CELL ORGANS AND SYSTEMS CONSTRUCTIONS.

CHANGING OF ONE CELL SPECIES TO OTHER, UNDER CELL PIS DECISIONS, DIFFERENTIATION OF THE CELLS, CONSTRUCTIONS OF VARIETY OF DIFFERENT SPECIES, ALL CHALLENGED BY PIS.

Genesis of Dominant Live and Non Live Molecules and biomolecules inside cells, and cell function, all are under Cell's Particle Intelligence System center (PIS) orders. The Biomolecules also receive Particle Clouds and emit

Particle Cloud orders from self into PCS.
The RNA, DNA, Genes and other internal cell Bio-Molecule constructions, done under A. S. I. Mic. U. CIC. through PIS orders.

Present time's, RNA Genesis, DNA Genesis, Cell Genesis, Cell and Tissue Replication systems Genesis all are Equal and are Copies of past Bio Molecules and Molecules Evolution Era Pathways.

Today's Internal Cell RNA Genesis, DNA Genesis, internal Cell's Organ Genesis, internal Cell's System Genesis are Equal to past Evolution's Cell Neo Genesis, RNA – DNA Neo Genesis.

THIRD PHASE EVOLUTION IN EARTH, OR EVOLUTION IN MACRO UNITS

Molecular Evolution And Living Thing Genesis

Nano Unit Genesis + Cell Genesis + Organs & Systems Genesis = Earth's Molecular Evolution = genesis of living things

---

PARTICLE INTELLIGENCE SYSTEM CENTERS OF LIVING THING NANO UNITS, MICRO UNITS, BODY ORGAN - PIS, BODY SYSTEM - PIS, AND CNS – PIS, ALL WITH CLOSE COORDINATION AND COOPERATION OF EACH OTHER DISCOVERED, AND CONSTRUCTED DIFFERENT SPECIES BODY ORGANS, AND BODY SYSTEMS CONSTRUCTIONS, UNDER THE A. S. I. Mol. CIC. AND THROUGH PARTICLE CLOUD ORDERS, UNDER MOLECULAR EVOLUTION LAWS AND RULES, CONSTRUCTED A LIVING SPECIES FROM CHEMICAL COMBINATIONS OF THE ORDINARY EXISTING ELEMENTS AND PARTICLES EXISTED AT EARTH.

TODAY'S EMBRYONIC STAGE STABLE MUTATIONS SEQUENCES, WHEN THESE SEQUENTIAL MUTATIONS ALTERNATE WITH EMBRYONIC STEM CELL'S MEDIA CHEMICAL FORMULARY CHANGE. ALL OF THESE PHENOMENONS ARE EQUAL AND THE SAME AS IT TOOK PLACE DURING THE PAST MOLECULES EVOLUTION ERAS.

ABOVE MENTIONED MULTI PARTICLE INTELLIGENCE SYSTEM CENTERS (PIS) UNDER THE MOLECULAR EVOLUTION ORDERS AND PATHS, RECONSTRUCT LIVING THING SPECIES MOLECULAR CONSTRUCTIONS EXACTLY UNDER SAME ORDER AS OCCURRED DURING PAST EVOLUTION ERA PATHS AND ORDERS, UNDER PIS PARTICLE CLOUD ORDERS, PCS, AND IN EMBRYO PRODUCE ANOTHER EXACT LIVING THING COPY FROM ORIGINAL SPECIES IN EMBRYONIC STATES.

EMBRYONIC STAGE LIVING THING GENESIS IS EQUAL TO EVOLUTION ERA LIVING THING GENESIS, UNDER PIS - CLOUD ORDERS & PCS.

TODAY DURING THE BODY TURN OVER PROCESS AND PHENOMENON, THE PARTICLE INTELLIGENCE SYSTEM CENTERS (PIS) OF CELLS AND TISSUES, BODY ORGAN'S PIS, BODY SYSTEM'S PIS, CNS – PIS, THE BIOMOLECULES AND GENES PIS, UNDER WELL DEFINED COORDINATION AND COOPERATION WITH EACH OTHER CONSTRUCT THE LIVING THINGS DIFFERENT BODY ORGANS AND SYSTEMS UNDER EXACT ORDERS AND EXACT COPIES OF PAST MOLECULAR EVOLUTION, IN ANY GIVEN ONE HALF LIFE TIME FRAMES SPECIFIC FOR THAT GIVEN LIVING THING SPECIES. THROUGH FUNDAMENTAL PARTICLE PRECISION ACCURACIES UNDER THE A. S. I. Mol. CIC WITH NO MISTAKES, AND CONSTRUCT ENORMOUS DIFFERENT LIVING THING SPECIES OF PLANTS, ANIMALS, INSECTS, ANPHIBIANS, BIRDS, ETC. IN ALL TIMES NON STOP, YEARS AFTER YEARS AUTONOMOUSLY THROUGH EXPONENTIAL SPEEDS. ABOVE IS PHENOMENON OF TURN OVER PROCESS.

Today's Cell and Tissue genesis, Organ and Living thing Genesis is copy of Molecular Evolution.

Autonomous Maintenances of Living Things under CNS – PIS and Evolution Orders

==

Autonomously Maintaining Construction of Genes, Bio Molecules, RNA, DNA, Cells, Body organs and systems, under A. S. I.

Mol. CIC. through CNS - PIS, O- PIS, S- PIS, PCS, P. Cl. Orders.

THE PARTICLE INTELLIGENCE SYSTEM CENTERS (PIS) OF ATOMS, PIS OF CELLS, THE PIS OF LIVING THINGS, CNS - PIS IN ORDER TO PRESERVE AND MAINTAIN THE EVOLUTION DISCOVERED AND MANUFACTURED LIVING THING BIO MACHINES, TISSUES, BODY ORGANS, BODY SYSTEMS, THE CELLS, THE INTERNAL CELL ORGANS AND SYSTEMS, INTERNAL CELL ATOM CONSTRUCTIONS, INTERNAL CELL PROTON CONSTRUCTIONS, INTERNAL CELL ELECTRON AND NEUTRON PARTICLE COMPOUND CONSTRUCTIONS. ETC. HAD TO BUILD ANOTHER DIFFERENT KIND BODY ORGANS AND SYSTEMS, IN ORDER TO MAINTAIN THE EVOLUTION MANUFACTURED LIVING THING CONSTRUCTION DISCOVERIES AS FOLLOWING:

THE CNS – PIS AND OTHER CELL AND ATOM PIS JOINTLY HAD TO INVENT, AND MANUFACTURE DIFFERENT OTHER KIND NEEDED ORGANS, SYSTEMS SUCH AS MUSCULOSKELETAL SYSTEM, ETC. IN ORDER THESE ORGANS TO ENABLE THE MOVEMENT CAPABILITY FOR LIVING THING SPECIES, TO SEARCH AND FIND NEEDED FOOD MOLECULES AND NEEDED FUNDAMENTAL PARTICLES FOR INTERNAL CELL AND INTERNAL ATOM CHEMICAL LAB.S USAGES, TO BUILD NEW CELLS, AND NEW ATOMS, DUPLICATE LIVING THING GENESIS, THROUGH USAGES OF THOSE EXOGENOUS PARTICLES AND ELEMENTAL MOLECULES.

THE MOLECULAR EVOLUTION CONSTRUCTED LIVING THING SPECIES CNS – PIS, BODY ORGAN – PIS, CELL AND ATOMS – PIS. DURING MOLECULAR EVOLUTION PARTICLE INTELLIGENCE SYSTEM CENTERS HAD TO WORK, DISCOVER AND CONSTRUCT OTHER NEEDED BODY SYSTEMS SUCH AS CARDIOVASCULAR

PUMPING BIO DEVICES, IN ORDER THIS CIRCULATION SYSTEM BIO DVICES PROVIDE NEEDED EXOGENOUS FOOD MOLECULES, EXOGENOUS FUNDAMENTAL PARTICLES, AND CELLS TO NEEDED LOCATION AT WHOLE BODY. THESE CELL – MOLECULE CIRCULATION SYSTEMS PROVIDE NEEDED MOLECULES TO CONSTRUCT ANOTHER DUPLICATE CELLS, THROUGH PROVIDING PARTICLES CONSTRUCT INTERNAL ELECTRON NUCLEON PARTICLE COMPOUND NEO GENESIS. ATOM GENESIS, CELL GENESIS, BIOMOLECULAR DUPLICATIONS INSIDE CELLS, ATOMS NEW PARTICLE COMPOUND CONSTRUCTIONS, NEO TISSUE GENESIS, AND CONSTRUCTION OF DUPLICATE CELLS AND CAUSE BODY GROWTH.

THE PARTICLE INTELLIGENCE SYSTEM CENTERS OF NANO UNITS, MICRO UNITS, MACRO UNITS, CNS – PIS, IN ORDER TO PRESERVE THEIR EVOLUTION INVENTED AND CONSTRUCTED LIVING THINGS STRUCTURES, HAD TO PROVIDE NEEDED MOLECULES AND BIOMOLECULES TO THEIR CHEMICAL LAB.S, IN ORDER THOSE MOLECULES TO BE USED FOR MANUFACTURING THE LIVINGT THINGS NEW RNA, DNA, GENE, CELL SYSTEMS AND CELL ORGANS, TISSUES AND ORGANS NEW CONSTRUCTIONS, UNDER A. S. I. Mol. CIC.

FOR ABOVE MENTIONED REASON, THE PIS HAD TO DISCOVER AND BUILD OTHER NEEDED ORGANS AND SYSTEMS, SUCH AS: G. I. S., G. U. S., LIVER, SPLEEN, PANCREAS, KIDNEYS, LUNGS, BONE MARROW, THORACIC – PULMONARY SYSTEMS, ETC. IN ORDER THESE ORGANS INSIDE THEIR CHEMICAL LAB.S UNDER A. S. I. Mol. CIC. TO CHANGE, AND REFINE ENVIRONMENTAL OUT SIDE'S RAW MATERIALS AND NON USABLE MOLECULES, INTO REFINED MOLECULES AND USABLE NEW CHEMICAL FORMULARY MOLECULES AND BIOMOLECULES.

WHICH THE LIVING THINGS CHEMICAL LAB.S, THEREAFTER REFING WERE CAPABLE TO USE THOSE REFINED OUTSIDE ORIGIN

MOLECULES, INSIDE THEIR CHEMICAL LAB.S AND CONSTRUCT DUPLICATE BIOMOLECULES, GENES, CELLS, TISSUES AND DUPLICATE LIVING THING UNDER A. S. I. Mol. CIC. ALSO CONSTRUCT NEW ENZYMES, HORMONES, REPLICA GENES, CHROMOSOMES, DUPLICATE RNA & DNA MOLECULES, CELL ORGANS AND CELL SYSTEMS REPLICATIONS, CELL GENESIS, TISSUE GENESIS, BODY ORGANS AND BODY SYSTEMS NEW CONSTRUCTIONS, AND LIVING THING GENESIS.

THE PIS OF LIVING THINGS IN ORDER TO PRESERVE AND MAINTAIN CONSTRUCTION OF INTERNAL ATOM PARTICLE COMPOUND CONSTRUCTIONS OF LIVING THINGS CNS ELECTRONS AND NUCLEONS AND AUTONOMOUS CENTRAL INTELLIGENCE SYSTEMS INTERNAL ATOM PARTICLE CLOUD COMPOUND CONSTRUCTIONS. THE PIS CONSTRUCTED DIFFERENT PERIPHERAL SENSORY ORGANS, SUCH AS LIGHT PARTICLE DETECTING SENSORY ORGANS, SONIC PARTICLE DETECTING SENSORY ORGANS, THERMAL PARTICLE DETECTING SENSORY ORGANS, ETC.

THE LIVING THING - PIS IN ORDER TO CONTROL CONTINUOUSLY THE DIFFERENT BODY ORGANS AND SYSTEMS CELLS AND ATOMS FUNCTION STATUS HAD TO CONSTRUCT AUTONOMOUS CENTERAL INTELLIGENCE SYSTEM CENTERS, AND PERIPHERAL PIS CONSTRUCTION.

THE ALL OF ABOVE DIFFERENT CONSTRUCTED PIS ARE IN NEEDS OF ENVIRONMENTAL PARTICLE CLOUDS, S. Y. E. T. – FP. – I.I. – P. Cl. AND ENVIRONMENTAL INFORMATION IMAGE PARTICLE CLOUDS, IN ORDER TO USE THESE EXOGENOUS PARTICLE CLOUDS CONSTRUCTION AS A REFERENCE INFORMATION FOR DECISION MAKING, ALSO THROUGH USAGE OF THOSE PARTICLE CLOUDS CONSTRUCT THEIR INTERNAL ELECTRON, NUCLEON AND CNS – ATOM'S PARTICLE CLOUD COMPOUND CONSTRUCTIONS (WHICH THESE ARE INFORMATION IMAGE CLOUD STORAGE SYSTEMS,

AND INTERNAL CNS ATOM'S SCIENCE AND KNOWLEDGE ACCUMULATION SYSTEMS). THIS IS PHENOMENON OF PSYCHE GENESIS, AND THOUGH CURRENT GENESIS.

IN ORDER TO COLLECT EXOGENOUS NEEDED FUNDAMENTAL PARTICLE INFORMATION AND IMAGE PARTICLE CLOUDS (F.P. – I.I. – P. Cl.), AND COMBINE THOSE PARTICLE CLOUDS WITH CNS INTERNAL ELECTRON NUCLEON PARTICLE COMPOUNDS, AND PRODUCE STORAGE OF PARTICLE CLOUD COMPOUNDS INSIDE CNS – ATOMS (STORGAE OF KNOWLEDGE AND SCIENCE PROCESSES INSIDE BRAIN). THE PIS CONSTRUCTED DIFFERENT SENSORY ORGANS, WHICH THESE SENSORY ORGANS FUNCTIONS ARE TO ATTRACT DIFFERENT KIND AIRBORNE FUNDAMENTAL PARTICLE SUCH AS LIGHT PARTICLE CLOUDS, SONIC PARTICLE CLOUDS, OR THERMAL PARTICLE CLOUDS, AND TRANSIT THOSE CAPTURED PARTICLE CLOUDS THROUGH PCS INTO CNS – ATOM'S INTERNAL ELECTRON AND INTERNAL NUCLEON CHEMICAL LAB.S AND UNDER A. S. I. FP. Mol. CIC. PRODUCE PARTICLE CLOUD COMPOUNDS, AND STORE EXOGENOUS ENVIRONMENTAL INFORMATION AND IMAGES IN PARTICLE CLOUD COMPOUND FORMS INSIDE THE CNS – ATOMS.

THESE ARE PHENOMENONS OF CNS - NEO PARTICLE CLOUD COMPOUND GENESIS, NEO ELECTRON GENESIS, NEO NUCLEON CONSTRUCTION, PSYCHE GENESIS PHENOMENONS AS WELL.

FOR COLLECTING OUTSIDE PARTICLE CLOUDS, THE PIS HAD TO CONSTRUCT OTHER BODY ORGANS AND SYSTEMS, SUCH AS CONSTRUCTION OF DIFFERENT LIGHT PARTICLE ATTRACTING ORGANS AND SYSTEMS (EYES) OR SONIC PARTICLE SENSORY ORGANS (AUDITARY ORGANS) AND THERMAL PARTICLE SENSORY ORGANS AND SENSORY SYSTEMS, LUNGS, CUTANEOUS SENSORY SYSTEMS (SENSE TEMPERATURE, SENSE PRESSURE, SENSE PARTICLE VIBRATION), GIS, ETC. IN ORDER THESE ORGANS ATTRACT, CAPTURE AND TRANSIT DIFFERENT KIND

ENVIRONMENTAL PARTICLES AND PARTICLE CLOUDS, INCLUDING S. Y. T. E. FP. – I.I. – P. Cl. AND FUNUNDAMENTAL PARTICLES, AND TRANSIT THESE PARTICLE CLOUDS INTO INTERNAL CNS ELECTRONS AND NUCLEONS CHEMICAL LAB.S, IN ORDER THESE PARTICLES TO BE USED IN CONSTRUCTIONS OF THE CNS CELLS INTERNAL ATOM PARTICLE COMPOUND CONSTRUCTIONS.

THESE PARTICLE CLOUDS ARE USED IN CONSTRUCTIONS OF INTERNAL ELECTRONS, INTERNAL NUCLEON PARTICLE CLOUD COMPOUND CONSTRUCTIONS, THE A. S. I. FP. Mol. CIC. INSIDE CNS ELECTRONS NUCLEONS BETWEEN THESE PARTICLE CLOUDS PRODUCE PSYCHE GENESIS, THOUGHT CURRENT GENESIS.

THESE ARE STORAGE SYSTEMS OF KNOWLEDGE, INFORMATION, IMAGE, INTELLIGENCE INSIDE THE CNS ATOMS CONSTRUCTIONS, IN PARTICLE CLOUD COMPOUND FORMS. THE CNS ELECTRONS NUCLEONS OVER THESE ACCUMULATED INFORMATIONS AND IMAGES CONSTRUCTED ENCYCLOPEDIA, INSIDE ELECTRONS NUCLEONS HAS TO COMPARE AND MATCH INCOMING EXTERNAL PARTICLE CLOUDS WITH EXISTING INTERNAL ELECTRON NUCLEON PARTICLE COMPOUND MODALITIES AND STANDARD CLOUD NORMS, UNDER A. S. I. FP. Mol. CIC. THEREAFTER THE THOROUGH ANALISIS, MAKE THEIR BEST POSSIBLE DECIONS, AND ISSUE PARTICLE CLOUD ORDERS, AND CONTROL ALL TOTAL BODY CELLS, TOTAL BODY NANO POPULATIONS PHYSICAL CHEMICAL BIOLOGICAL FUNCTIONS.

PARTICLE INTELLIGENCE SYSTEM CENTERS (PIS) DURING MOLECULAR EVOLUTION WORKED HARD AND DISCOVERED ENDOCRINE ORGANS, WHICH THESE ENDOCRINE SYSTEMS CHEMICAL LAB.S FROM ENVIRONMENTAL ORDINARY ELEMENTS AND PARTICLES, THROUGH A. S. I. Mol. CIC. & A. S. I. FP. Mol. CIC. CONSTRUCTED DOMINANT DIFFERENT INDEPENDENT MOLECULES (HORMONES), WHICH THE EMITTED PARTICLE CLOUD ORDERS FROM THESE DOMINANT MOLECULES, ARE

CAPABLE TO ORDER TO OTHER BODY ORGANS AND BODY SYSTEMS PARTICLE INTELLIGENCE SYSTEM CENTERS THAT, HOW THESE BODY ORGANS PIS MUST FUNCTION ACCORDING THE EXACT ORDERS OF DOMINANT MOLECULE'S PARTICLE CLOUD ORDERS.

UNDER THESE DOMINANT MOLECULES PARTICLE CLOUD ORDERS, THE RECIPIENT BODY SYSTEM'S ATOMS, CELLS, AND TISSUES PIS MUST FUNCTION EXACTLY THE SAME AS HAS BEEN ORDERED ACCORDING HORMONE MOLECULES PARTICLE CLOUD ORDERS. THE RECIPIENT BODY ORGANS HAS TO SUBMIT THEMDELVES, TO DO PHYSICAL, CHEMICAL BIOLOGICAL FUNCTIONS UNDER THE DOMINANT MOLECULES (HORMONE) PARTICLE CLOUD ORDERS, EXACTLY AS IT HAS BEEN ORDERED TO THEM, HOW TO CONSTRUCT RNA, DNA, CHROMOSOMES, INTERNAL CELL ORGANS AND SYSTEMS, THE CELL CONSTRUCTIONS AND CELL FUNCTIONS, TISSUES, ORGANS CONSTRUCTIONS AND FUNCTIONS.

THE RECIPIENT BODY ORGANS ELECTRONS, NUCLEONS HAD TO CONSTRUCT THEIR INTERNAL ATOM ELECTRON, NEUTRON, PROTON PARTICLE COMPOUND CONSTRUCTIONS ACCORDING THE DOMINANT HORMONE MOLECULES CLOUD ORDERS. ALSO THE INTERNAL CELL ORGANS, SYSTEMS, AND CHEMICAL LAB.S OF RECIPIENT ORGANS CELLS AND ATOMS MUST WORK BIOLOGICALLY, PHYSICALLY, CHEMICALLY ONLY UNDER AND ACCORDING PARTICLE CLOUD ORDERS. THE DOMINAT MOLECULES PIS CLOUD ORDERS ALWAYS HAVE EXACT COORDINATION, AND COOPERATION WITH CNS – PIS PARTICLE CLOUDS ORDERS.

TODAYS, ONGOING BODY ORGANS AND SYSTEM GENESIS AND

FUNCTIONS, ARE EQUAL TO THE MOLECULAR EVOLUTION'S ORGAN AND SYSTEM GENESIS AND FUNCTIONS, EXACTLY THE SAME.

The Fundamental Particles sources for Electron and Nucleon Genesis

Bio molecule and Molecule sources for Cell Genesis in Living Things

CONSTRUCTIONS OF LIVING THINGS ORGANS, UNDER DOMINANT MOLECULE PARTICLE CLOUD ORDERS

THE DOMINANT MOLECULES SUCH AS HORMONES EMIT ORDER PARTICLE CLOUD FROM SELF, THE ORDER PARTICLE CLOUDS CONTROL PARTICLE INTELLIGENCE SYSTEMS OF RECIPIENT ORGANS. THE FUNDAMENTAL PARTICLES AND PARTICLE CLOUDS POSSESSES DIFFERENT KIND PARTICLE CLOUD SPECIFIC THERMAL, ELECTRIC, SONIC, LIGHT ENERGY CONTENT, PROVIDING THESE DIFFERENT KIND QUANTUM ENERGIES TO RECIPIENT ORGANS BY DOMINANT MOLECULES, ALSO CAN TRIGGER START DIFFERENT TYPE

CHEMICAL, PHYSICAL, BIOLOGICAL AND MECHANICAL TASKS.

PROVIDING DOMINANT MOLECULES PARTICLE CLOUD ORDERS TO DIFFERENT RECIPIENT ORGANS PARTICLE INTELLIGENCE SYSTEM CENTERS, CAN TRIGGER START PHYSIOLOGICAL, CHEMICAL, PHYSICAL FUNCTIONS IN RECIPIENT ORGANS AND PRODUCE CONSTRUCTION OF GENES, BIOMOLECULES, CELLS, TISSUES, OR TRIGGER START PHYSICAL MECHANICAL BIOLOGICAL FUNCTIONS. STOPPING PARTICLE CLOUD DELIVERIES BY DOMINANT MOLECULES TO RECIPIENT ORGANS PIS CENTERS, ALSO CAN STOP ONGOING ORGAN TASKS OF RECIPIENT ORGAN'S UNDER ORGANS PIS CONTROL.

IN EARTH THE ELECTRONS, NUCLEONS, ATOMS, AND NANO UNITS, THAT CONSTRUCTING THE LIVING THINGS CELLS, BODY TISSUES, DOMINANT MOLECULES, ETC. HAVE BEEN CONSTRUCTED FROM ORDINARY ELEMENTS AND FUNDAMENTAL PARTICLES WHICH EXPLAINED IN THIS BOOK (AND NOTHING ELSE).

UNDER A. S. I. Mol. CIC. IODINE IS ESSENTIAL TO BUILD DOMINANT (HORMONE) MOLECULES, OR IRON IS ESSENTIAL TO CONSTRUCT SOME KIND VITAL CELLS. OR CALCIUM MOLECULES THROUGH A. S. I. Mol. CIC. CONSTRUCT BONE CELLS, ETC. THESE ARE THE WAYS THE EXISTING FUNDAMENTAL PARTICLES AND EXISTING EARTH'S ELEMENTS THROUGH A. S. I. Mol. CIC. UNDER EVOLUTION ORDERS CONSTRUCTED LIVING THINGS CELLS, TISSUES, ORGANS, ATOMS AND MOLECULES. AND OPERATE THE

LIVING THINGS BIOLOGICAL, PHYSICAL, CHEMICAL FUNCTIONS.

{[ENDOCRINE ORGAN'S INTERNAL CELL CHEMICAL LAB.S, AND INTERNAL ATOM CHEMICAL LAB.S, UNDER A. S. I. Mol. CIC. AND A. S. I. FP. Mol. CIC. FROM CHEMICAL COMBINATION OF ORDINARY ELEMENTAL MOLECULES, AND FUNDAMENTAL PARTICLE MOLECULES WITH EACH OTHER PRODUCE DOMINANT HORMONE MOLECULES, AND EXCRETE CONSTRUCTED DOMINANT MOLECULES (HORMONES) INTO BLOOD CIRCULATION SYSTEMS. THE HORMONES TRAVEL TO DIFFERENT TARGET BODY ORGANS AND BODY SYSTEMS. IN RECIPIENT BODY ORGANS. THE EMITTED PARTICLE CLOUDS FROM DOMINANT MOLECULES, JOINTLY WITH CNS – PIS EMITTED PARTICLE CLOUDS ALTOGETHER, CONTROL AND OPERATE FUNCTIONS OF THE RECIPIENT ORGAN'S - PIS, RECIPIENT ORGANS CELL'S - PIS, AND RECIPIENT ORGANS NANO UNITS -PIS FUNCTION OPERATIONS. AND THESE DIFFERENT PARTICLE INTELLIGENCE SYSTEM CENTERS ARE IN CHARGES OF OPERATIONS OF INTERNAL – CELLS, INTERNAL ELECTRONS, NUCLEONS, INTERNAL BODY ORGANS TISSUES PHYSICAL, CHEMICAL, BIOLOGICAL FUNCTIONS. IN TARGET ORGANS, THE DOMINANT MOLECULES (HORMONES) THROUGH PARTICLE CLOUD EMISSION FROM INTERNAL MOLECULAR PIS, PCS, OPERATE INTERNAL ELECTRON, NUCLEON AND INTERNAL CELL A. S. I. Mol. CIC. FUNCTIONS. AND CAUSE PRODUCTION OF LIVE AND NON LIVE DIFFERENT MOLECULE CONSTRUCTIONS IN RECIPIENT CELLS, ATOMS, NANO UNITS. THE DOMINANT MOLECULES CAUSE BIO MOLECULE GENESIS, CELL'S MOLECULE DUPLICATION, GENE AND

CHROMOSOME GENESIS AT CELL'S CHEMICAL LAB.S. CONSEQUENTLY PRODUCE BODY TISSUES, CELLS, TISSUE AND BODY GROWTH, LIVING THING GENESIS. ALL OF THESE FUNCTIONS CONTROLLED AND ORDERED UNDER DOMINANT MOLECULES PIS, PCS, AND HIERARCHY SYSTEMS.

ALSO THE EMITTED PARTICLE CLOUDS FROM CNS – PIS, AND PARTICLE CLOUD EMISSION FROM DOMINANT MOLECULES JOINTLY BOTH CONTROL, OPERATE AND CONSTRUCT INTERNAL ELECTRON NUCLEON PARTICLE COMPOUND CONSTRUCTIONS. CONSTRUCT NEW ELECTRONS - NUCLEONS, (PHENOMENON OF NEO ELECTRON GENESIS WITH PARTICLE CLOUDS, NEO NUCLEON GENESIS, PSYCHE- GENESIS PROCESS OF CNS).

THE LIVING THING GENESIS, UNDER CNS - PIS AND COMMANDER DOMINANT MOLECULE'S PIS JOINTLY OPERATED BY PARTICLE CLOUD ORDERS, CELL GENESIS, ATOM GENESIS UNDER A. S. I. FP. Mol. CIC. ALL WHICH ARE GOING ON TODAY, ARE EQUAL TO PAST MOLECULAR EVOLUTION ORDERS, OPERATIONS AND CONSTRUCTIONS.

THE LIVING THING GENESIS OF TODAY ARE EXACT PATHS AND ORDERS OF MOLECULAR EVOLUTION OF PAST. THE ORDERS AND PATHS OF PAST MOLECULE EVOLUTION IS EQUAL TO TISSUE GENESIS OF TODAY, THESE EVOLUTION ORDERS ARE DIFFERENT AT DIFFERENT SPECIES. BUT THESE EVOLUTION ORDERS AND PATHS ALWAYS ARE CONSTANT AND RELATIVELY FIX FOR EACH GIVEN LIVING THING SPECIES]}.

MALFUNCTIONING OF PARTICLE INTELLIGENCE SYSTEMS (PIS)

===============================
====

And Normal Functioning - PIS

MALFUNCTIONING O – PIS, MALFUNCTIONING CNS – PIS, MALFUNCTIONING OF DOMINANT MOLECULES – PIS, MALFUNCTIONING PARTICLE CLOUD CIRCULATION SYSTEMS, MALFUNCTIONING CELL'S PIS, MALFUNCTIONING ELECTRON'S PIS, MALFUNCTIONING NEUTRON'S PIS, MALFUNCTIONING PROTON'S PIS, MALFUNCTIOING ATOM'S PIS, MALFUNCTIONING PARTICLE CLOUDS, MALFUNCTIONING INFORMATION AND IMAGE PARTICLE CLOUDS SYSTEM.

Normal Functioning PIS and Particle Clouds

THE DOMINANT CNS ELECTRON – PIS (E – PIS) CENTERS, DOMINANT CNS NUCLEON - PIS (N – PIS) CENTERS, AND DOMINANT CNS ATOM'S – PIS, THROUGH A. S. I. FP. Mol. CIC. PRODUCE AND EMIT PARTICLE CLOUD ORDERS. THE PRODUCED PARTICLE CLOUDS FROM DIFFERENT CNS – PIS,

THROUGH PARTICLE CLOUD CIRCULATION SYSTEMS TRAVEL INTO DIFFERENT PERIPHERAL ORGAN PIS CENTERS.

UNDER HIERARCHY MOLECULAR ORDER OPERATION SYETEMS THE CNS – PIS PARTICLE CLOUDS CONTROL, OPERATE, AND ORDER DIRECTLY TO SUBORDINATE ORGAN PARTICLE INTELLIGENCE SYSTEM CENTER (O – PIS) HOW TO PERFORM THE ORGAN'S ENTIRE PHYSICAL, CHEMICAL BIOLOGICAL TASK OPERATIONS ACCORDING THE CNS – PIS CLOUD ORDERS.

ADDITIONALLY, DOMINANT MOLECULE – PIS, JOINTLY WITH CNS – PIS, THROUGH PARTICLE CLOUD ORDER OPERATION SYSTEMS ORDER UPON O – PIS. ALTOGETHER WITH WELL DEFINED COORDINATION AND COOPERATION THROUGH PARTICLE PRECISION ACCURACIES, WITH EXPONENTIAL SPEEDS, OPERATE, CONTROL, AND ACHIEVE ALL BODY ORGAN'S BIOLOGICAL AND CHEMICAL FUNCTIONS. THE FUNCTIONS OF ENTIRE ORGAN'S TOTAL POPULATION ELECTRONS, ATOMS, NEUTRONS, PROTONS ALL ACHIEVED UNDER HIERARCHY SYSTEMS OF DOMINANT MOLECULES PIS, O- PIS, CNS – PIS.

THE CNS – PIS, DOMINANT MOLECULE - PIS, AND O- PIS ALL JOINTLY UNDER PARTICLE CLOUD HIERARCHY SYSTEMS OPERATE, MANAGE, CONTROL AND ACHIEVE THE TOTAL BODY CELL POPULATION'S INDIVIDUAL CELL'S FUNCTIONS, BODY TISSUE FUNCTIONS, BODY ORGAN'S FUNCTIONS, AND BODY SYSTEM'S PHYSICAL, CHEMICAL, BIOLOGICAL FUNCTION OPERATIONS.

THE ENTIRE PERIPHERAL ORGAN'S ELECTRONS, NUCLEONS, ATOMS, CELL'S TOTAL POPULATION'S - PIS, ALL ARE

SUBORDINATE (PIS), IN COMPARE TO CNS – PIS, AND DOMINANT MOLECULE'S - PIS.

THE PERIPHERAL SUBORDINATE PIS CENTERS HAVE ONLY LIMITED, RELATIVE AUTONOMY TO OPERATE AND FUNCTION INDEPENDENTLY. THE SUBORDINATE PIS ACHIEVE THEIR PHYSICAL, CHEMICAL, BIOLOGICAL FUNCTIONS UNDER ISSUED PARTICLE CLOUD ORDERS FROM THEIR SUPERIOR PIS CENTERS. AND ACCORDING THE SUPERIOR CNS - PIS ORDER PARTICLE CLOUDS, WHICH HAS BEEN EMITTED BY CNS- PIS, AND DOMINANT MOLECULE – PIS ORIGIN.

NORMAL & ABNORMAL PARTICLE CLOUDS

Normal and Abnormal CNS - internal Atom's Particle Cloud Compound Constructions

NORMAL PARTICLE CLOUD GENESIS FROM NORMAL CNS - PIS

THE NORMAL PARTICLE CLOUD COMPOUND CONSTRUCTED CNS – CENTER ELECTRONS, NUCLEONS, ATOMS UNDER NUMEROUS DIFFERENT REGENERATIVE AND DEGENERATIVE A. S. I. FP. Mol. CIC. PRODUCE NORMAL PARTICLE CLOUDS, AND EMIT INTO PARTICLE CIRCULATION SYSTEMS.

THE NORMAL CONSTRUCTION PARTICLE CLOUDS ORDERS TRAVEL INTO RECIPIENT SUBORDINATE ORGAN'S PARTICLE INTELLIGENCE SYSTEM CENTERS (O-PIS). AND CNS – PIS EMITTED PARTICLE CLOUD ORDERS, JOINTLY WITH O- PIS PRODUCE NORMAL FUNCTION FOR ORGAN'S TOTAL [ELECTRON, NEUTRON, PROTON, ATOM, CELL AND TISSUES] POPULATION, AT RECIPIENT SUBORDINATE ORGANS.

THE PHYSICAL CHEMICAL BIOLOGICAL FUNCTIONS IN RECIPIENT SUBORDINATE ORGAN'S ELECTRONS, NUCLEONS, ATOMS, CELLS, AND TISSUE FUNCTIONS, ALL DONE ACCORDING THE PARTICLE CLOUD ORDERS OF SUPERIOR CNS – CENTER'S ATOMS DECISIONS, UNDER HEIRARCHY SYSTEMS. AND PRODUCE NORMAL FUNCTION OPERATION AT DIFFERENT RECIPIENT SUBORDINATE PERIPHERAL BODY ORGANS.

ABNORMAL PARTICLE CLOUD GENESIS FROM ABNORMAL CNS - PIS

THE ABNORMAL FUNDAMENTAL PARTICLE INFORMATION AND IMAGE PARTICLE CLOUD COMPOUND CONSTRUCTED CNS – ELECTRONS, PROTONS, NEUTRONS, UNDER DEGENERATIVE A. S. I. FP. Mol. CIC. AT DIFFERENT CNS –

CENTERS RELEASE ABNORMAL PARTICLE CLOUDS INTO CNS – PCS, OR INTO GENERAL BODY – PCS, AND TRAVEL INTO DIFFERENT PERIPHERAL ORGANS. THIS PHENOMENON IS RETRIEVAL OF ABNORMAL PARTICLE CLOUDS FROM CONSTRUCTED ABNORMAL PARTICLE CLOUD INFORMATION AND IMAGE PARTICLE CLOUD COMPOUNDS, IN DIFFERENT CNS CENTERS.

THE RELEASED ABNORMAL PARTICLE CLOUDS CURRENTS WHEN CIRCULATE BETWEEN DIFFERENT CNS – PIS CENTERS AND INTERACT WITH EACH OTHER UNDER REGENERATIVE OR DEGENERATIVE A. S. I. FP. Mol. CIC.

THESE ABNORMAL PARTICLE CLOUD - PCS INSIDE DIFFERENT CNS CENTERS, MAY SENSE AS ABNORMAL THOUGHT CURRENTS, OR ABNORMAL PSYCHE. PRODUCE THOUGHT DISORDERS OF DIFFERENT TYPES.

THE RELEASED ABNORMAL PARTICLE CLOUDS CURRENTS FROM ABNORMAL CNS – PIS, ALSO CAN TRANSIT INTO PERIPHERAL RECIPIENT ORGAN'S PIS CENTER, ABNORMAL PARTICLE CLOUD ORDERS CAN PRODUCE NON FUNCTIONING, OR MALFUNCTIONAL ORGANS OF ENORMOUS DIFFERENT PATTERNS. PARALYSIS OF NORMAL EXTREMITY UNDER ABNORMAL PARTICLE CLOUD ORDERS IS AN EXAMPLE IN ELPILEPSY, OR OTHER ENORMOUS DIFFERENT PATTERN DISORDERS IN DIFFERENT RECIPIENT ORGANS.

THE ABNORMAL ORDER PARTICLE CLOUDS THROUGH GENERAL PARTICLE CIRCULATION SYSTEMS (PCS) TRANSIT INTO DIFFERENT SUBORDINATE RECIPIENT BODY ORGANS PARTICLE INTELLIGENCE SYSTEM CENTERS (O – PIS). THE

RECIPIENT O – PIS UNDER ORDERS OF ABNORMAL PARTICLE CLOUDS CAN PRODUCE ABNORMAL FUNCTIONING RECIPIENT ORGAN'S ELECTRONS, NUCLEONS, ATOMS, CELLS AND TISSUE IN ENORMOUS DIFFERENT PATTERNS IN DIFFERENT BODY SYSTEMS.

CONTROLS OF TOTAL BODY ATOM POPULATION FUNCTIONS BY CNS- PIS

"CNS – PIS" Diseases

DISEASES OF "PARTICLE INTELLIGENCE SYSTEMS"

Particle Cloud Produced Diseases

Cloud Disorders of Dominant Molecules

Particle Cloud Emission and PCS Diseases

THE CNS ELECTRONS – NUCLEONS THROUGH A. S. I. FP. Mol. CIC. AND UNDER DOMINANT CNS ELECTRONS AND NUCLEONS PARTICLE INTELLIGENCE SYSTEM CENTERS (CNS – PIS) DECIDE, ORDERS, PRODUCE AND EMIT ORDER PARTICLE CLOUDS FROM SELF INTO BODY PCS.

THE SUBORDINATE PERIPHERAL BODY ORGANS OF LIVING THINGS UNDER ATTRACTIVE GRAVITON FORCE CAPTURE THESE CNS – PIS ORDER PARTICLE CLOUDS FROM GENERAL BODY PARTICLE CIRCULATION SYSTEMS (PCS). UNDER THE ELECTRON – NUCLEON HIERACHY ORDER SYSTEMS, THE PERIPHERAL RECIPIENT SUBORDINATE ORGAN'S PARTICLE INTELLIGENCE SYSTEM CENTERS (O – PIS) HAS TO FUNCTION UNDER EXACT ORDERS OF THESE DOMINANT CNS – PIS PARTILCE CLOUD ORDERS.

WHEN A NORMAL ELECTRON NUCLEON CONSTRUCTION – CNS, UNDER A. S. I. FP. Mol. CIC. PRODUCE AND EMIT NORMAL PARTICLE CLOUD FROM SELF INTO BODY'S GENERAL PARTICLE CIRCULATION SYSTEMS (PCS), THE RECIPIENT PERIPHERAL ORGAN'S PARTICLE INTELLIGENCE SYSTEM CENTERS (O – PIS) MUST FUNCTION NORMALLY, EXACTLY AS ORDERED BY NORMAL DOMINANT CNS – PIS. THAT MEANS THE RECIPIENT BODY ORGAN'S ELECTRONS, NUCLEONS, ATOMS, TISSUES, AND CELLS MUST FUNCTION PHYSICALLY, CHEMICALLY, BIOLOGICALLY NORMAL AS THEY HAVE BEEN ORDERED AND INSTRUCTED ACCORDING CNS – PIS PARTICLE CLOUDS. DURING LIFE LONG THESE FUNCTIONS ALSO CHECKED AND CONTROLLED BY CNS - PIS. UNLESS SOME FACTORS DISTURB NORMAL PARTICLE CLOUD OPERATION SYSTEMS.

ABNORMAL CONSTRUCTION NANO UNITS IN CNS- PIS

IN PATHOLOGICAL SITUATIONS UNDER ANY REASON, WHEN THE CNS ELECTRON –NUCLEON STRUCTURES HAVE BEEN CONSTRUCTED THROUGH USES OF ABNORMAL FUNDAMENTAL PARTICLE MOLECULES INFORMATION AND IMAGE PARTICLE CLOUD COMPOUND CONSTRUCTIONS. IN THESE SITUATIONS, THE ABNORMAL S. Y. T. E. – FP. I.I. P. Cl. – COMPOUND CONSTRUCTION IS ALSO EQUAL TO ABNORMAL PSYCHE AND THOUGHT CURRENT PARTICLE CLOUD STORAGE PHENOMENON INSIDE THE CNS ELECTRONS AND NUCLEONS.

(ABNORMAL CNS – PIS PARTICLE CLOUD COMPOUND CONSTRUCTION PHENOMENON IS EQUAL TO ABNORMAL PSYCHE GENESIS OF DIFFERENT KINDS PSYCHIATRIC DISEASES, DEPENDING TO THE PARTICLE CLOUD KINDS).

THESE ABNORMAL CNS – NANO UNITS CONSTRUCTIONS UNDER REVERSE A. S. I. FP. Mol. CIC. RELEASE ABNORMAL PARTICLES AND ABNORMAL PARTICLE CLOUDS (S. Y. T. E. – FP – I.I. – P. Cl.) TO OUT OF MOLECULAR COMPOUND CONSTRUCTIONS CNS ATOMS, ELECTRONS AND NUCLEONS CONSTRUCTIONS AS FREE PARTICLE CLOUDS CIRCULATE INTO CNS PARTICLE CIRCULATION SYSTEMS OF THE BRAINS DIFFERENT CENTERS (PCS). THIS IS RETRIEVAL PHENOMENON OF PARTICLE CLOUDS, WHICH IT IS SAME AS PARTICLE CLOUD EMISSION FROM CNS ELECTRONS AND NUCLEONS, ALSO IN LIVING THINGS THIS PHENOMENON PRODUCE BIOLOGICAL SIGNS AND SYMPTOMS WHICH BY PUBLIC AS CALLED MEMORIZATION PROCESS OR PHENOMENON.

IN ABNORMAL CNS - ELECTRON, NUCLEON PARTICLE CLOUD (Y. T. E. S. – FP – I.I. – P. Cl. –COMPOUND) CONSTRUCTION CASES OF THE BRAIN CELLS, THE EXOGENOUS ABNORMAL S. Y. T. E. – FP. –

I.I. – P. Cl. ENTER INTO CNS ELECTRONS, NUCLEONS AND COMBINE WITH INTERNAL ATOM MOLECULES AND PRODUCE ABONORMAL PARTICLE COMPOUND MOLECULE CONSTRUCTIONS OF THE CNS ELECTRONS AND NUCLEONS, UNDER DEGENERATIVE - REGENERATIVE A. S. I. FP. Mol. CIC.

DURING THE RETRIEVAL PHENOMENON, THE PRODUCED AND EMITTED FREE ABNORMAL SONIC, LIGHT, THERMAL, ELECTRIC FUNDAMENTAL PARTICLE INFORMATION AND IMAGE PARTICLE CLOUDS (S. Y. T. E. – FP. – I.I. – P. Cl.) ENTER INTO GENERAL BODY PARTICLE CIRCULATION SYSTEMS (PCS) AND TRAVEL ALL OVER TO DIFFERENT RECIPIENT PERIPHERAL BODY ORGANS AND BODY SYSTEMS.

IN THE PERIPHERAL BODY ORGANS AND SYSTEMS, THESE ABNORMAL PARTICLE ORDERS CLOUDS UNDER A. S. I. FP. Mol. CIC. INTERACT WITH SUBORDINATE BODY ORGAN'S PARTICLE INTELLIGENCE SYSTEM CENTERS (PIS) AND CAUSE PRODUCTIONS OF THE ABNORMAL BODY ORGAN FUNCTIONS, ABNORMAL BODY SYSTEM FUNCTIONS, DEPENDING WHICH CNS- PIS CENTERS AND WHICH PERIPHERAL ORGANS PCS CONNECTIONS ARE INVOLVED.

THIS PHENOMENON IS THE ESSENTIAL CAUSES OF THE PRESENTLY KNOWN NEUROTIC DISEASES (NEUROSIS), AS WELL AS IS THE CAUSES FOR ENORMOUS ABNORMAL OTHER PATHOLOGICAL CONDITIONS WHICH ARE KNOWN UNDER IDIOPATHIC DISEASES, AUTOIMMUNE DISEASES, OTHER MEDICAL PROBLEMS.

IN PERIPHERAL BODY ORGANS, THE RECIPIENT SUBORDINATE ORGAN'S PARTICLE INTELLIGENCE SYSTEM CENTERS (O – PIS) MUST FUNCTION ABNORMALLY EXACTLY ACCORDING THE ORDERED CNS - PIS ABNORMAL ORDER CLOUDS, UNDER COMMANDING HIERARCHY SYSTEMS OF THE CNS – PIS ELECTRONS AND NUCLEONS ORDERS.

THE SUBORDINATE O – PIS MUST PERFORM AND WORK EXACTLY AS ORDERED BY CNS – PIS PARTICLE CLOUDS ORDERS. THE RESULT OF THESE HIERARCHY NANO UNIT PARTICLE CLOUD OPERATION IS THE RECIPIENT ORGAN MUST PERFORM ABNORMAL A. S. I. FP. Mol. CIC., ABNORMAL PHYSICAL MECHANICAL FUNCTIONS, ABNORMAL CHEMICAL AND BIOLOGICAL FUNCTIONS, UNDER ABNORMAL CNS – PIS PARTICLE CLOUD ORDERS. THE PATIENT'S SENSORY AND MOTOR ORGANS MANIFESTATIONS REVEAL AND EXPLAIN THE KIND OF ABNORMALITY EXIST IN PARITCLE CLOUD ABNORMALITIES.

THESE ARE DISEASES OF THE PARTICLE CLOUDS, DISEASES OF THE PARTICLE INTELLIGENCE SYSTEM CENTERS, AND DISEASES OF THE PARTICLE CIRCULATION SYSTEMS (PCS). DISEASES OT THE ELECTRONS, PROTONS, NEUTRONS, ATOMS AND MCIRO UNIT CONSTRUCTIONS. THESE DISORDERS PRODUCED BY SICK ATOMS, SICK ELECTRONS AND NUCLEONS. NORMAL CONSTRUCTION ATOMS PRODUCE NORMAL PSYCHE, NORMAL THOUGHT CURRENTS.

LAB. TESTS, MRI, XRAY, SONOGRAM, PET SCAN CAN NOT DETECT PIS, PCS, P. Cl. ABNORMALITY. THESE DISEASES ARE INTERNAL ATOM DEFECTS.

The Symptoms and Signs of PIS, PCS, P. Cl. Diseases

The internal Electron and internal Nucleon Diseases symptoms are similar to Cells and tissue diseases symptoms

INTERNAL ELECTRON - NUCLEON DISEASES

THE ABNORMAL CONSTRUCTION INTERNAL ELECTRON - NUCLEON FUNDAMENTAL PARTICLE COMPOUNDS AT THE BRAIN'S INVOLUNTARY AND VOLUNTARY CNS - PARTICLE INTELLIGENCE SYSTEMS CENTERS (CNS – PIS) PRODUCE ABNORMAL SIGNS AND SYMPTOMS BOTH PHYSICAL AS WELL AS FUNCTIONAL (PSYCHE) TYPES.

THE MAL -CONSTRUCTED NON VOLUNTARY AND VOLUNTARY CNS CENTERS SICK ELECTRONS - NUCLEONS HAVE BEEN CONSTRUCTED THROUGH USES OF ABNORMAL SONIC, THERMAL, ELECTRIC AND LIGHT INFORMATION AND IMAGE PARTICLE CLOUD COMPOUND CONSTRUCTIONS UNDER A. S. I. FP. Mol. CIC. THESE SICK ATOMS, ELECTRONS, NUCLEON'S PARTICLE INTELLIGENCE SYSTEM CENTERS DECIDE, PRODUCE AND EMIT ABNORMAL PARTICLE CLOUDS AND RELEASE THE PRODUCED PARTICLE CLOUDS INTO PARTICLE CIRCULATION SYSTEMS (PCS).

THE EMITTED ABNORMAL FUNDAMENTAL PARTICLE INFORMATION AND IMAGE PARTICLE CLOUDS (S. Y. T. E. – FP – I.I. – P. Cl.) THROUGH PCS TRAVEL TO DIFFERENT PERIPHERAL ORGANS, THE ABNORMAL PARTICLE CLOUDS IN RECIPIENT ORGANS PRODUCE MALFUNCTION ACCORDING ABNORMAL PARTICLE CLOUDS ORDERS BIOLOGICALLY, MECHANICALLY, CHEMICALLY AND PHYSICALLY, IN PERIPHERAL ORGANS.

THE ABNORMAL FUNCTIONS IN RECIPIENT ORGANS PRODUCE DEVELOPMENTS OF ABNORMAL SIGNS AND SYMPTOMS IN CLOUD RECIPIENT SUBORDINATE BODY SYSTEMS AND ORGANS. THE SICK ATOM, SICK PROTON, ELECTRON, NEUTRON PRODUCED SIGNS AND SYMPTOMS HAVE REMARKABLE SIMILARITIES TO SICK CELL DISEASES.

THIS IS BRIEFLY PATHOGENESIS OF SYMTOMS AND SIGNS OF SICK ELECTRONS AND NUCLEONS. THE PRODUCED SIGNS AND SYMPTOMS ARE ELECTRON – NUCLEON DISEASES SYMPTOMS. IN INTERNAL ATOM DISEASES THE PARA CLINICAL TESTS MOSTLY ARE NORMAL. NEED ADVANCED FUTURE TECHNOLOGIES TO DETECT INTERNAL ELECTRON NUCLEON MALFUNCTIONS AND CONSTRUCTION ABNORMALITIES.

Signs of sick CNS-Cardiac – PIS (C- PIS) diseases, Sick CNS- Cardiac PCS & Particle Cloud Disorders.

(Pathogenesis of Cardiac - CNS – PIS Electrons Nucleons diseases)

THE EXOGENOUS ENVIRONMENTAL PRODUCED INCOMING ABNORMAL S. Y. T. E. – FP – I.I. – P. Cl. (THE EXOGENOUS-SYMPTOM PRODUCING PARTICLE CLOUDS), OR THE INDIGENOUS PRODUCED ABNORMAL S. Y. T. E. – FP – I.I. – P. Cl. (INDIGENOUSLY PRODUCED ABNORMAL SYMPTOM PRODUCING - PARTICLE CLOUDS) SUCH AS ABNORMAL COUGH PRODUCTION PARTICLE CLOUDS, ABNORMAL BRONCHOSPASM PRODUCING PARTICLE CLOUDS, INCOMING ABNORMAL WHEEZING – GENESIS PARTICLE CLOUDS, OR ABNORMALLY PRODUCED INCOMING CHEST PAIN AND DYSPNEA PRODUCING PARTICLE CLOUDS TRAVEL TOWARD CNS CARDIO PULMONARY CENTERS ELECTRON NUCLEON'S SUBSYSTEM UNITS, VIA INTERNAL ATOM PARTICLE CIRCULATION SYSTEMS (A - PCS), AND ENTER INTO CNS – CARDIOPULMONARY CENTERS ELECTRONS AND NUCLEONS SUBSYSTEM CHEMICAL LAB.S.

THESE EXOGENOUS OR INDIGENOUSLY PRODUCED ABNORMAL INFORMATION SYMPTOM PARTICLE CLOUDS FROM DIFFERENT PIS CENTERS TRAVEL THROUGH GENERAL PCS FINALLY ENTER INTO CNS CARDIO PULMONARY ELECTRONS NUCLEONS CENTERS INTERNAL SUBSYSTEM UNIT'S CHEMICAL LAB.S, UNDER REGENERATIVE A. S. I. FP. Mol. CIC. COMBINE WITH CNS

INTERNAL ELECTRON NUCLEONS PARTICLE COMPOUND MOLECULES, PRODUCE ABNORMAL SYMPTOM PARTICLE CLOUD COMPOUND CONSTRUCTIONS AND PARTICLE CLOUD STORAGE SYSTEMS INSDIE CNS INTERNAL ATOM MOLECULAR CONSTRUCTIONS, STORE ABNORMAL PARTICLE CLOUDS INSIDE CNS – PIS CARDIO PULMONARY CENTERS ATOMS.

ABOVE PROCESS CONSTRUCT ABNORMAL PARTICLE CLOUD COMPOUND CONSTRUCTIONS AND ABNORMAL PARTICLE CLOUD STORAGE SYSTEMS INSIDE CARDIO PULMONARY CNS INTERNAL ATOMS CONSTRUCTIONS IN SICK PARTICLE CLOUD COMPOUND CONSTRUCTION FORMS. WHEN THESE ABNORAL CLOUDS THROUGH PARTICLE CIRCULATION SYSTEMS TRAVEL DIFFERENT BODY ORGANS AND BODY SYSTEMS, THE DIFFERENT BODY ORGANS AND SYSTEMS HAS TO FUNCTION ACCORDING THE DOMINANT CLOUD ORDERS, AND THE RELATED ABNORMAL SIGNS AND SYMPTOMS OF SICK PARTICLE CLOUDS PRODUCE THE EXACT ORGAN SIGNS AND SYMPTOMS OF CLOUDS, ALTHOUGH THE ORGANS AND BODY SYSTEMS PHYSICALLY ARE NOT SICK CONSTRUCTIONS.

THIS IS ABNORMAL SYMPTOM PARTICLE CLOUD STORAGE SYSTEM PHENOMENONS INSIDE CNS – PIS, AND INSIDE CNS CARDIO PULMONARY CENTERS ELECTRONS AND NUCLEONS, UNDER DEGENERATIVE A. S. I. FP. Mol. CIC. CAN BE RELEASED AS FREE ABNORMAL PARTICLE CLOUDS AND TRAVEL THROUGH DIFFERENT INDIGENOUS OR EXOGENOUS PARTICLE CIRCULATION SYSTEMS (PCS) ANY PLACE IN UNIVERSE.

THIS IS PATTERNS OF PARTICLE CLOUD TRANSMISSION AND PARTICLE CLOUD CONTAGION BETWEEN DIFFERENT SUBJECTS. THESE ARE ROUTES OF NORMAL OR ABNORMAL PARTICLE CLOUD TRANSMIT FROM ONE SUBJECT TO OTHER. THE RELATIVE CONDITIONAL REFLEXES IN REGARD TO PARTICLE CLOUDS

GENESIS HAS MAJOR ROLES IN PRODUCTION OF THIS PHENOMENON.

THE EXOGENOUS OR INDIGENOUSLY PRODUCED ABNORMAL SYMPTOM PARTICLE CLOUDS ENTER INTO VOLUNTARY AND INVOLUNTARY CNS CARDIO PULMONARY CENTERS OF RECIPIENT SUBJECTS, COMBINE WITH CNS INTERNAL ELECTRON – NUCLEON PARTICLE COMPOUNDS, UNDER REGENERATIVE A. S. I. FP. Mol. CIC. AND PRODUCE ABNORMAL SYMPTOM S. Y. T. E. – FP. I.I. – P. Cl. COMPOUND CONSTRUCTIONS INSIDE THE CNS CARDIO PULMONARY VOLUNTARY AND INVOLUNTARY CNS – CENETR'S ATOMS.

THIS IS PHENOMENON OF STORAGE ABNORMAL SYMPTOM PARTICLE CLOUD COMPOUNDS INSIDE CARDIO PULMONARY CNS CENTERS ELECTRONS AND NUCLEONS. ALSO THIS IS PROCESS OF ABNORMAL CNS ELECTRON NUCLEON SYMPTOM PARTICLE CLOUD CONSTRUCTIONS (GENESIS OF ABNORMAL PARTICLE CLOUD CONSTRUCTIONS INSIDE CNS – ATOMS, OR ABNORMAL ATOM GENESIS PHENOMENON). THESE ABNORMAL CONSTRUCTED ELECTRONS NUCLEONS UNDER REVERSE REGENERATIVE A. S. I. FP. Mol. CIC. RELEASES AND PRODUCE ABNORMAL SYMPTOM PARTICLE CLOUDS FREE FORMS, THE RELEASED ABNORMAL PARTICLE CLOUDS TRAVEL THROUGH PCS TO THE HEART AND LUNG ORGANS.

EMITTED ABNORMAL PARTICLE CLOUD ORDERS OF CARDIAC CNS - PIS CENTERS, TRAVEL TO CARDIAC PARTICLE INTELLIGENCE SYSTEM CENTER IN LEFT AURICLE, ALSO EMITTED ABNORMAL PARTICLE CLOUDS FROM THORACO PULMONARY CNS – PIS CENTERS TRAVEL TO LUNGS AND THORAX PARTICLE INTELLIGENCE SYSTEM CENTERS. THE LUNGS, THORAX, HEART HAS FUNCTION UNDER CNS HIERARCHY PARTICLE CLOUD ORDERS SYSTEMS, AND ACCORDING CNS CLOUD ORDERS, PRODUCE FUNCTION, SIGNS AND SYMPTOMS ACCORDING INCOMING

ABNORMAL OR NORMAL CARDIO PULMONARY CLOUD ORDERS FROM CNS - PIS.

THE ABNORMAL CARDIO PULMONARY FUNCTIONS AND RELATED SYMPTOMS AND SIGNS SUCH AS; DYSPNEA, COUGH, CHEST PAIN, BRONCHOSPASM, DYSRHYTHMIA, WHEEZING, AND OTHER ABNORMAL CARDIO PULMONARY FUNCTIONS AND RELATED SIGNS AND SYMPTOMS ARE PRODUCED UNDER DOMINANT ABNORMAL CNS PARTICLE CLOUD ORDERS.

THE ABNORMAL PARTICLE CLOUD ORDERS IN RECIPIENT LUNGS CAN PRODUCE DISEASES, SYMPTOMS AND SIGNS SIMILAR TO ASTHMA OR OTHER LUNG DISEASES, ALSO THE ABNORMAL PARTICLE CLOUD ORDERS IN HEART CAN PRODUCE CHEST PAIN, DYSRHYTHMIA, ALL OTHER ACUTE OR CHRONIC CARDIAC DISORDERS FROM MI TO DEATH AND ARREST AND OTHER COMPLICATIONS.

CNS – GIS - PIS center's Atom sicknesses,
CNS – GUS -PIS center's Atom Diseases

===========================

====

Signs of GIS-CNS Sick Electron, Sick Protons & Neutrons

Symptoms and Signs of abnormally constructed CNS-GUS Center's Sick Electron, Sick Nucleon Construction Disease

++
+++++++++

PARTICLE CLOUD DISEASES OF BODY ORGANS

THERE ARE TWO MAIN SOURCES FOR FUNDAMENTAL PARTICLE INFORMATION AND IMAGE PARTICLE CLOUDS, WHICH THESE TWO PARTICLE CLOUD SOURCES EMIT FP – I.I. – P. Cl., AND THESE PARTICLE CLOUDS TRAVEL AND ENTER INTO CNS ELECTRONS, PROTONS, NEUTRONS SUBSYSTEM UNIT'S CHEMICAL LAB.S. AND THESE PARTICLE CLOUDS THROUGH REGENERATIVE A. S. I. FP. Mol. CIC. COMBINE WITH INTERNAL ELECTRON, NUCLEON PARTICLE COMPOUNDS AND CONSTRUCT CNS INTERNAL ELECTRON NUCLEON S. Y. E. T. – FP – I.I. – P. Cl. (PARTICLE CLOUD)- COMPOUND CONSTRUCTIONS.

THE FIRST MAJOR PARTICLE CLOUD SOURCE IS FROM EXOGENOUS ENVIRONMENTAL SUBJECT'S EMITTED

FUNDAMENTAL PARTICLE INFORMATION AND IMAGE PARTICLE CLOUD SOURCES. WHICH OUTSIDE WORLD ENVIRONMENTAL SUBJECTS ALL EMIT EXOGENOUS FUNDAMENTAL PARTICLE INFORMATION AND IMAGE PARTICLE CLOUDS (EX - FP – I.I. – P. Cl.) FROM THEMSELVES. THE NORMAL SUBJECTS EMIT NORMAL PARTICLE CLOUDS AND ABNORMAL SUBJECTS EMIT ABNORMAL PARTICLE CLOUDS FROM SELF.

THE EXOGENOUS PARTICLE CLOUDS IN EARTH MOSTLY TRAVEL AIRBORN, THE DIFFERENT SENSORY ORGANS ATOMS OF THE LIVING THINGS, UNDER ATTRACTIVE GRAVTON FORCES OF THEIR ELECTRONS AND NUCLEONS ATTRACT AND CAPTURE THESE FLOATING AIRBORNE CLOUDS, AND TRANSIT THE CAPTURED PARTICLE CLOUDS INTO DIFFERENT CNS CENTERS ELECTRONS AND NUCLEONS. THE SONIC PARTICLE CLOUDS THROUGH SENSORY EXTERNAL HEARING ORGAN ATOM'S GRAVITON FORCES GET ATTRACTION, CAPTURED, TRANSITTED INTO AUDITARY CNS CENTERS INTERNAL ATOMS SUBSYSTEM UNITS CHEMICAL LAB.S.

CONTRARY THE LIGHT FUNDAMENTAL PARTICLE CLOUDS CAPTURED BY VISION (OPHTHALMIC) SENSORY SYSTEMS ATOMS ATTRACTIVE GRAVITON FORCES, AND TRANSIT INTO INTERNAL ELECTRON NUCLEON CHEMICAL LAB.S OF THE VISION CNS CENTER SUBSYSTEM UNITS CHEMICAL LAB.S, INSIDE THESE CHEMICAL LAB.S THE TRANSMITTED (NORMAL OR ABNORMAL S. Y. T. E. –FP – I.I. – P. Cl.) PARTICLE CLOUDS COMBINE WITH INTERNAL ELECTRON NUCLEON SUBSYSTEM UNITS PARTICLE COMPOUNDS, AND CONSTRUCT NORMAL OR ABNORMAL CNS CENTERS INTERNAL ELECTRON NUCLEON PARTICLE CLOUD COMPOUND CONSTRUCTIONS.

THESE ARE STORAGE SYSTEMS OF THE NORMAL PARTICLE CLOUDS, OR STORAGE SYSTEMS OF THE ABNORMAL PARTICLE

CLOUDES INSIDE THE CNS ELECTRONS, NUCLEONS, ATOMS IN PARTICLE CLOUD COMPOUND FORMS.

THE SECOND AND NEXT MAJOR SOURCES OF THE S. Y. E. T. – FP – I.I. –P. Cl. ARE INDIGENOUS SOURCES OF PARTICLE CLOUDS, WHICH THESE GROUP PARTICLE CLOUDS PRODUCED INSIDE BODY BY LIVING THING INTERNALLY. THE INDIGENOUS PARTICLE CLOUDS ALSO DIVIDE INTO TWO GROUPS OF NORMAL PARTICLE CLOUDS AND ABNORMAL PARTICLE CLOUDS. THE NORMAL PARTICLE CLOUDS CONSTRUCT NORMAL CONSTRUCTION INTERNAL ELECTRON NUCLEON PARTICLE CLOUD COMPOUND CONSTRUCTIONS AND ABNORMAL PARTICLE CLOUDS CONSTRUCT ABNORMAL CONSTRUCTION INS INTERNAL ATOM PARTICLE CLOUD COMPOUND CONSTRUCTIONS.

THE NORMALLY CONSTRUCTED CNS ELECTRONS NUCLEONS PRODUCE AND EMIT NORMAL PARTICLE CLOUDS, AND THROUGH HIERARCHY DOMINANT NORMAL PARTICLE CLOUD ORDERS IN PERIPHERAL BODY ORGANS AND SYSTEMS PRODUCE NORMAL BODY FUNCTIONS WITH NORMAL BODY SYSTEMS BEHAVIORS, NORMAL FEELING TYPES BODY SIGNS, AND NORMAL FEELING TYPES BODY SYMPTOMS.

THE ABNORMAL ELECTRON NUCLEON CONSTRUCTED CNS ATOMS, ELECTRONS AND NUCLEONS THROUGH THEIR ABNORMALLY CONSTRUCTED PARTICLE COMPOUND CONSTRUCTIONS AND ABNORMAL CONSTRUCTED PARTICLE CLOUDS, UNDER DOMINANT PARTICLE CLOUD HIERARCHY ORDER SYSTEMS, PRODUCE ABNORMAL BIOLOGICAL FUNCTIONS AND ABNORMAL BEHAVIOUR PRODUCTIONS AT PERIPHERAL SUBORDINATE BODY ORGANS AND BODY SYSTEMS. THE CREATED ABNORMAL FUNCTIONING AT PERIPHERAL ORGANS CELLS AND TISSUES CAUSE PRODUCTIONS AND FEELINGS OF ABNORMAL SIGNS AND SYMTOMS.

{[IN ALL BODY ORGANS AND SYSTEMS, THE NORMALLY CNS CONSTRUCTED PARTICLE CLOUD COMPOUND CONSTRUCTION CENTERS, UNDER A. S. I. FP. Mol. CIC PRODUCE NORMAL PARTICLE CLOUDS. THE EMITTED NORMAL PARTICLE CLOUDS FROM CNS ATOMS, ELECTRONS AND NUCLEONS CAUSE THE RECIPIENT PERIPHERAL SUBORDINATE BODY ORGANS FUNCTION BIOLOGICALLY, CHEMICALLY, PHYSICALLY AND MECHANICALLY NORMAL. AND PRODUCE NORMAL SENSING BODY SIGNS AND BODY SYMPTOMS (NORMAL FEELING LIVING THINGS) IN ENTIRE PERIPHERAL SUBORDINATE ORGANS UNDER DOMINANT CLOUD ORDER SYSTEMS.

IN CONTRARY, THE ABNORMALLY CONSTRUCTED CNS PARTICLE COMPOUND ELECTRONS, NUCLEONS AND ATOMS, WHICH HAVE BEEN CONSTRUCTED UNDER ABNORMALLY TRANSMITTED EXOGENOUS OR INDIGENOUS S. Y. T. E. – FP – I.I. – P. Cl. CONSTRUCTIONS, AND HAVE PRODUCED ABNORMALLY CONSTRUCTED EXOGENOUS AND INDIGENOUS S. Y. T. E. – FP – I.I. – P. Cl. COMPOUND CNS - ATOM CONSTRUCTIONS, UNDER A. S. I. FP. Mol. CIC., THESE ABNORMALLY CONSTRUCTED PARTICLE CLOUD COMPOUNDS, UNDER DEGENERATIVE A. S. I. FP. Mol. CIC. PRODUCE, AND EMIT FROM THEMSELVES ABNORMAL S. Y. T. E. – FP – I.I. – P. Cl. CLOUDS. THESE EMITTED ABNORMAL PARTICLE CLOUDS, VIA PCS TRAVEL INTO PERIPHERAL ORGAN'S O – PIS, AND BODY SYSTEM'S – PIS. SUCH AS; GUS - PIS, GIS -PIS, MUSCULO SKELETAL - PIS, CARDIO PULMONARY -PIS, CUTANEOUS – PIS, ETC.

THESE TRANSMITTED ABNORMAL S. Y. E. T. – FP- I.I. – P. Cl. UNDER THEIR DOMINANT HIERACHY PARTICLE CLOUD ORDER SYSTEMS, IN DIFFERENT RECIPIENT ORGANS SUCH AS GASTRO - INTENSTINAL SYSTEMS, GENITO URINARY SYSTEMS, MUSCULO SKELETAL SYSTEMS, CUTANEOUS SYSTEMS, ETC. CAUSE

PRODUCTION OF ABNORMAL ORGAN FUNCTIONING AND ABNORMAL BODY SYSTEM FUNCTIONING UNDER CLOUD ORDERS.

THE BODY ORGAN'S MALFUNCTIONS AND BODY SYSTEM'S MALFUNCTIONS, UNDER ABNORMAL PARTICLE CLOUD ORDERS, SECONDARILY PRODUCE ABNORMAL BODY ORGAN SIGNS AND ABNORMAL BODY SYSTEM SYMPTOMS. WHICH ALL ARE PRODUCED FALSELY EXACTLY AS ORDERED BY ABNORMAL CNS PARTICLE CLOUD ORDERS.

THE GIS PARTICLE CLOUD DISORDERS CAUSE PRODUCTION OF ABDOMINAL PAINS, GASTRO INTESTINAL MALFUNCTIONS AND THEIR RELATED ABNORMAL SIGNS AND SYMPTOMS, ALL ALSO CAN CAUSE CHRONIC COMPLICATIONS STARTING FROM INFECTIONS, INFLAMATIONS, UP TO MALIGANACY DEGENERATIVE CHANGES IN CHRONIC STATES.

THE GUS PARTICLE CLOUD DISEASES MAY PRODUCE FREQUENCY OR RETENSION DISORDERS, DISRUPTION OF NORMAL FUNCTION AND PRODUCTION OF RELATED ABNORMAL SIGNS AND SYMPTOMS. AND ALL BODY ORGANS ARE AFFECTED UNDER PARTICLE CLOUD ORDER DISEASES, AND ELECTRON NUCLEON SICKNESSES. WHICH THEIR DESCRIPTIONS ARE SEVERAL TEXT BOOK SIZES]}.

THESE ARE PARTICLE CLOUD DISEASES. THESE ARE ATOM SICKNESSES. THESE ARE ELECTRON NUCLEON MAL-CONSTRUCTION DISORDERS. ELECTRON NUCLEON MALFUNCTIONING DISEASES. THE PRESENTLY AVAILABLE PARA CLINICAL TESTS ARE NEGATIVE FOR THESE DISEASES. THE TECHNOLOGY FOR DIAGNOSING THE ELECTRON NUCLEON LEVEL DISORDERS HAVE NOT BEEN DISCOVERED PRESENTLY.

SICK CNS - ATOMS
The Abnormal Particle Cloud Construction of CNS Electron –Neutron - Proton

The Psychiatric diseases are Particle Cloud Transmitted Diseases
(Mental Disorders are Contagious)

Healthy CNS Atoms
The Normal Particle Cloud Compound Construction CNS Electrons - Nucleons — Atoms

THE NORMAL OR ABNORMALLY ACCUMULATED INFORMATION IMAGE CLOUDS INSIDE CNS ATOMS

Normal Psyche Genesis
HEALTHY ATOMS

THE NORMAL FUNDAMENTAL PARTICLE INFORMATION IMAGE PARTICLE CLOUD COMPOUND CONSTRUCTED CNS ELECTRONS - NUCLEONS ARE NORMAL (HEALTHY) ATOMS, THE NORMAL PARTICLE CLOUD COMPOUND CONSTRUCTION HEALTHY ATOMS UNDER DEGENERATIVE A. S. I. FP. Mol. CIC. PRODUCE, EMIT AND RELEASE NORMAL S. Y. T. E. FP –I.I. – P. Cl. (PARTICLE CLOUD) FROM SELF INTO EXOGENOUS AS WELL AS INTO INDIGENOUS PARTICLE CIRCULATION SYSTEMS (PCS).

THE EMITTED PARTICLE CLOUDS FROM NORMAL CNS ELECTRONS AND NUCLEONS TRAVEL THROUGH EXOGENOUS PARTICLE CLOUD CIRCULATION SYSTEMS (EX. PCS) AIRBORNE INTO OTHER INDIVIDUALS. THE SENSORY ORGAN OF OTHER INDIVIDUALS THROUGH THEIR'S ATTRACTIVE ATOMS GRAVITON FORCES CAPTURE AIRBORNE PARTICLE CLOUDS FROM SPACE, AND TRANSIT THOSE PARTICLE CLOUDS INTO RECIPIENT PEOPLE'S INTERNAL CNS ELECTRONS NUCLEONS, AT DIFFERENT PARTICLE CLOUD RELATED BRAIN CENTERS (SONIC PARTICLES TRANSIT INTO AUDITARY CNS CENTERS, THE LIGHT PARTICLE CLOUD TRANSIT INTO VISION CENTER CNS, ETC.).

AT INSIDE CNS INTERNAL SUBSYSTEM UNITS CHEMICAL LAB.S OF RECIPIENT CNS ELECTRONS AND NUCLEONS, THESE ARRIVING NORMAL (S. Y. T. E. FP. I.I. P. Cl.) PARTICLE CLOUDS COMBINE WITH INTERNAL ELECTRON NUCLEON PARTICLE COMPOUNDS AND PRODUCE NORMAL CONSTRUCTION PARTICLE CLOUD COMPOUNDS (S. Y. T. E. FP. –I.I. – P. Cl. COMP.), UNDER A. S. I. FP. Mol. CIC., THESE ARE STORAGE SYSTEMS OF ENVIRONMENTAL INFORMATIONS AND IMAGE DOCUMENTS STORED IN PARTICLE CLOUDS COMPOUND FORMS, INSIDE THE CNS ELECTRONS AND NUCLEONS.

THIS PHENOMENON IS ENVIRONMENTAL SCIENCE AND KNOWLEDGE RECORDING AND STORAGE SYSTEMS IN PARTICLE CLOUD COMPOUND FORMS INSIDE THE CNS ELECTRONS AND

NUCLEONS. WHICH IT START FROM FIRST MINUTES AFTER THE BIRTH AND CONTINUE DURING THE ALL LIFE. THESE ARE ESSENTIAL FOR LEARNING, TEACHING AND TRAINING (IT IS CLOUD STORAGE SYSTEM INSIDE CNS ATOMS).

THE HEALTHY ATOMS THROUGH REGENERATIVE A. S. I. FP. Mol. CIC. PRODUCE AND EMIT FROM SELF NORMAL PARTICLE CLOUDS. THE NORMAL CNS - ATOMS EMIT AND TRANSIT FROM SELF NORMAL PARTICLE CLOUDS TO OTHER INDIVIDUALS. NORMAL PARTICLE CLOUDS IN RECIPIENT INDIVIDUALS CAUSE CONSTRUCTIONS OF NORMAL PARTICLE CLOUD COMPOUND CONSTRUCTION ELECTRONS AND NUCLEONS.

THE CHILDREN AND STUDENTS OF NORMAL PARENTS AND NORMAL TEACHERS CONSTRUCT NORMAL CNS PARTICLE CLOUD COMPOUN CONSTRUCTION ATOMS. THE NORMAL CONSTRUCTION ATOMS ARE NORMAL PARTICLE CLOUD DONORS.

THIS PHENOMENON IS NORMAL PSYCHE GENESIS AND NORMAL THOUGHT CURRENTS GENESIS, THE PROCEDURE OF NORMAL CONSTRUCTION CNS ELECTRONS AND NUCLEONS IN RECIPIENTS CHILDREN, STUDENTS AND SUBJECTS. ABOVE ARE ESSENTIAL OF NORMAL KNOWLEDGE ACCUMULATION SYSTEMS AND NORMAL INTELLIGENCE SYSTEM GENESIS PROCEDURES, INSIDE CHILDREN AND STUDENTS CNS ELECTRONS, NEUTRONS, AND PROTONS.

Sick Atoms

THE ABNORMAL CNS PARTICLE CLOUD (S. Y. T. E. –FP – I.I. – P. Cl.) CONSTRUCTION ELECTRON NUCLEON PARTICLE COMPOUND STRUCTURES, PRODUCE AND EMIT ABNORMAL PARTICLE CLOUDS (S. Y. T. E. – FP – I.I. – P. Cl.) FROM THEIR ABNORMALLY CONSTRUCTED CNS ATOMS INTO ENVIRONMENT AND INTO EXOGENOUS AND INDIGENOUS PARTICLE CIRCULATION SYSTEMS.

THE ABNORMALLY PRODUCED INFORMATION AND IMAGE PARTICLE CLOUDS TRAVEL AIRBORNE AND ENTER INTO RECIPIENT PEOPLES (SUCH AS KIDS, STUDENTS, SOCIAL AND POLITICAL GROUPS, ETC.) CNS INTERNAL ELECTRON AND INTERNAL NUCLEONS SUBSYSTEM UNITS CHEMICAL LAB.S.

THE INCOMING ABNORMAL PARTICLE CLOUDS (S. Y. T. E. –FP – I.I.- P. Cl.) FROM ABNORMAL DONOR ORIGINS (SUCH AS FROM ABNORMAL PARTICLE CLOUD DONOR PARENTS AT HOME) ENTER INTO RECIPIENTS CNS – INTERNAL ELECTRON NUCLEON SUBSYSTEM UNITS CHEMICAL LAB.S, AND UNDER A. S. I. FP. Mol. CIC. COMBINE WITH RECIPIENTS (FOR EXAMPLE THE CHILDREN AT HOME) INTERNAL ELECTRON NUCLEON PARTICLE COMPOUND MOLECULES, AND PRODUCE CONSTRUCTION OF ABNORMAL PARTICLE CLOUD (S. Y. T. E. – FP – I.I. – P. Cl.) COMPOUND MOLECULES INSIDE RECIPIENTS CNS ELECTROMS AND NUCLEONS (FOR EXAMPLE INSIDE THE CHILDRENS CNS ATOMS). THE ABNORMAL PARTICLE CLOUD DONORS CONSTRUCT ABNORMAL PARTICLE CLOUD CONSTRUCTION ATOMS (SICK ATOM) IN RECIPIENTS.

TRANSMISSION OF ABNORMAL PARTICLE CLOUDS (MENTAL DISORDERS) FROM ABNORMAL PARTICLE CLOUD CONSTRUCTION CNS SICK ATOM DONORS, TO NORMAL ATOM CONSTRUCTION

AND HEALTHY PARTICLE CLOUD CONSTRUCTION CNS ELECTRONS AND NUCLEONS INDIVIODUALS MOSTLY ARE TIME CONSUMING, IN CHRONIC CASES. BUT ABNORMAL PARTICLE CLOUD TRANSMISSIONS ALSO CAN OCCUR FAST AND ACUTELY IN SHORT TIMES IN MANY TIMES AS WELL.

A BIOLOGICAL CLASSIFICATION FROM PARTICLE CLOUDS

1 - THE CLASSES OF NORMAL FUNDAMENTAL PARTICLE CLOUDS (N-FP-Cl.):

THESE PARTICLE CLOUDS CONSTRUCT FUNDAMENTAL PARTICLE CLOUD COMPOUND MOLECULAR CONSTRUCTIONS OF AVERAGE ORDINARY NORMAL DIFFERENT LIVING THINGS SPECIES POPULATION INTERNAL BARAIN PARTICLE COMPOUND STRUCTURES OF INTERNAL ELECTRON NUCLEON MOLECULAR COMPOUNDS CONSTRUCTION OF ALL DIFFERENT PLANTS AND ANIMALS LIVING THING SPECIES OF PLANET EARTH.

THE NORMAL STANDARD ORDINARY CNS – PIS ATOMS HAVE BEEN CONSTRUCTED FROM NORMAL INFORMATION AND IMAGES PARTICLE CLOUD COMPOUND CONSTRUCTIONS. THE NORMAL CNS – PIS, O – PIS, S- PIS ELECTRONS, NEUTRONS, PROTONS AND ATOM'S INFORMATION AND IMAGE PARTICLE CLOUD COMPOUNDS CONSTRUCTIONS UNDER REG. DEG. A. S. I. FP. Mol. CIC. DECIDE, PRODUCE AND EMIT NORMAL PARTICLE CLOUNDS INTO GENERAL PARTICLE CIRCULATION SYSTEMS (PCS) FROM CNS – PIS CENTERS.

FREE PARTICLE CLOUDS CIRCULATE IN NORMAL PCS BETWEEN NORMAL DIFFERENT ORGANS AND TOTAL BODY ELECTRONS NUCLEONS PIS CENTERS POPULATIONS, THESE PIS, P. Cl., PCS ARE

IN INCHARGE OF NORMAL BODY PSYCHE AND PHYSICAL CHEMICAL FUNCTION OPERATIONS. NORMAL PIS, NORMAL PCS, NORMAL PARTICLE CLOUD PRODUCE NORMAL PSYCHE AND NORMAL PHYSICAL CHEMICAL AND BIOLOGICAL OPERATIONS AND FUNCTIONS FOR PLANTS AND ANIMAL LIVING THINGS.

2 – THE ABNORMALLY CONSTRUCTED CNE ELECTRONS NUCLEONS PARTICLE CLOUD COMPOUND CONSTRUCTIONS HAVE BEEN MADE FROM ABNORMAL TYPES INFORMATIONS AND IMAGES CLOUD COMPOUNDS, THE CNS – PIS WHEN DECIDE, PRODUCE AND EMITTED THESE FUNDAMENTAL ABNORMAL PARTICLE CLOUDS FROM CNS – PIS INTO ORDINARY PCS. THESE ABNORMAL PARTICLE CLOUDS IN PERIPHERAL ATOMS, ELECTRONS, NUCLEONS PARTICLE INTELLIGENCE SYSTEMS CAUSE MALFUNCTIONS, ABNORMAL FUNCTIONS PHYSICALLY AND CHEMICALLY BIOLOGICALLY IN INFLICTED BODY ORGANS AND SYSTEMS. PRODUCE AND CAUSE ABNORMAL SIGNS AND SYMPTOMS MANIFESTATIONS PHYSICALLY AS WELL AS PSYCHOLOGICALLY. WHICH ARE NOT AVERAGE STANDARD NROMAL LEVELS AND LIMITS OF FUNCTIONS.

SICK ATOMS

PSYCHIATRIC DISEASES ARE PARTICLE CLOUD TRANSMISSION DISEASES

Particle Cloud Diseases Are Contagious

THE ABNORMAL PARTICLE CLOUD CLASSES

1 -DEPRESSION PARTICLE CLOUDS (DEPRESSION S. Y. E. T. – FP – I.I. – P. Cl.): TRANSMIT THROUGH AIRBORNE EXOGENOUS PARTICLE CLOUD CIRCULATION SYSTEMS BETWEEN INDIVIDUALS (EX. PCS). THE CNS ELECTRONS, NEUTRONS, PROTONS OF THESE INFECTED PATIENTS MOSTLY HAVE BEEN CONSTRUCTED FROM DEPRESSION - PARTICLE CLOUDS (S. Y. E. T. – FP – I.I. – P. Cl.) COMPOUND CONSTRUCTIONS, THE CNS ATOMS OF THESE PATIENTS PRODUCE AND EMIT DEPRESSION PARTICLE CLOUDS AND TRANSIT DEPRESSION PARTICLE CLOUDS BACK AN FORTH BETWEEN INFECTED AND NON INFECTED INDIVIDUALS.

2 – PSYCHOSES PARTICLE CLOUDS. (PSYCHOSIS S. Y. E. T. – FP – I.I. – P. Cl.): THE CNS ELECTRONS NEUTRONS PROTONS OF THE PSYCHOSIS CLOUD INFECTED INDIVIDUALS ALSO PRODUCE AND EMITS PSYCHOSIS PARTICLE CLOUDS OF ENORMOUSLY DIFFERENT KINDS FROM CNS ATOMS. THE PSYCHOSIS PARTICLE CLOUDS ARE DIFFERENT TYPES IN DIFFERENT PATIENTS AND ARE CONTAGIOUS. THE CNS ATOMS PARTICLE COMPOUND CONSTRUCTIONS OF THIS SUBCLASS PATIENTS (SUCH AS SCHIZOPHRENIA) HAVE BEEN CONSTRUCTED MOSTLY FROM PSYCHOSIS INFORMATIONS AND IMAGE PARTICLE CLOUDS COMPOUND CONSTRUCTIONS (PSYCHOSIS S. Y. T. E. – FP – I.I. – P. Cl.). THE CNS ATOMS OF THIS SUBCLASS PATIENTS PRODUCE AND EMIT PSYCHOSIS PARTICLE CLOUDS.

THE EMITTED PSYCHOSIS INFORMATION AND IMAGE PARTICLE CLOUDS TRAVEL THROUGH BOTH INDIGENOUS –PCS AS WELL AS THROUGH EXOGENOUS – PCS BETWEEN THE NORMAL AND INFECTED PARTICLE CLOUD INDIVIDUALS. THE AIRBORNE PSYCHOSIS PARTICLE CLOUDS TRAVEL FROM INFECTED PATIENTS TO NORMAL INDIVIDUALS PRODUCE PSYCHOTIC PARTICLE CLOUD TRANSMITTED DISEASE IN RECIPIENT INDIVIDUALS.

3- BIPOLAR PARTICLE CLOUDS (BIPOLAR S. Y. T. E. – FP – I.I. – P. Cl.) CLASS: SIMILARLY, THE BIPOLAR PARTICLE CLOUDS ARE

CONTAGIOUS, THESE CLOUDS TRANSMIT FROM ONE PERSON TO OTHER, TRANSMISSION OF THESE PARTICLE CLOUDS PRODUCE BIPOLAR CLOUD INFECTED CNS – ATOMS OF BIPOLAR PATIENTS, OR BIPOLAR INFECTION DISEASES. THESE PARTICLE CLOUDS PRODUCE BIPOLAR PARTICLE CLOUD COMPOUND CONSTRUCTION IN CNS ELECTRONS NUCLEONS OF CONTAMINATED PATIENTS.

4 – NEUROSIS PARTICLE CLOUDS (NEUROSIS S. Y. T. E. – FP – I.I. – P. Cl.): TRANSMISSION AND GETTING INFECTED IS SIMILAR TO ABOVE MENTIONED ABOVE DISEASES. NEUROSIS PATICLE CLOUD TRANSMISSION PRODUCE NEUROTIC PATIENTS.

5 – DIFFERENT DEVIATION PARTICLE CLOUDS (DEVIATION S. Y. E. T. – FP – I.I. – P. Cl.): THESE PARTICLE CLOUDS INFLICTIONS ARE CAUSE OF ENORMOUS DIFFERENT TYPES SOCIAL, POLITICAL, MENTAL DEVIATION DISORDERS AND SICKNESSES. 6 - DIFFERENT CRIMINAL CLASSES OF PARTICLE CLOUDS (Y. S. T. E. – FP – I.I. – P. Cl.): THESE PARTICLE CLOUDS CONSTRUCT THE INFLECTED CRIMINALS INTERNAL CNS ELECTRON NUCLEON PARTICLE CLOUD COMPOUND CONSTRUCTIONS, THE EMITTED PARTICLE CLOUDS FROM SICK INDIVIDUALS ARE CONTAGIOUS, AND EMITTED PARTICLE CLOUDS TRANSMIT FROM INFECTED INDIVIDUALS TO OTHERS, THESE PARTICLE CLOUDS MAKE INFLICTED INDIVIDUALS TO COMMIT DIFFERENT PATTERNS CRIMES AND MUDERS.

(Normal - PCS, Abnormal - PCS) Fundamental Particle Cloud Circulation Systems

PARTICLE CIRCULATION SYSTEMS DISEASES

(MALFUNCTIONING - PCS)

========================

INDIGENOUS - PCS, AND EXOGENOUS- PCS

THE FUNDAMENTAL PARTICLE CIRCULATION SYSTEM (PCS), AND PARTICLE CLOUD CIRCULATION SYSTEMS DIVIDE INTO TWO MAJOR CLASS:

THE EXOGENOUS FUNDAMENTAL PARTICLE CIRCULATION SYSTEMS (EX. – PCS) AND EXOGENOUS FUNDAMENTAL PARTICLE CLOUD CIRCULATIONS SYSTEMS ARE BOTH THE EXTERNAL ATOM AIRBORNE OR EXTERNAL ATOMS PERIPHERAL ATOM FUNDAMENTAL PARTICLE CURRENTS AND EXOGENOUS PATICLE CLOUD CIRCULATION SYSTEMS.

THE FUNDAMENTAL PARTICLES AND PARTICLE CLOUDS CAN TRANSIT IN PERIPHERAL ATOM SPACE PARTICLE CIRCULATION SYSTEMS (PAS – PCS) IN DIFFERENT MEDIAS SUCH AS LIQUIDS, PLASMA, SOLIDS, ETC, AND PERIPHERAL ATOM SPACE PARTICLE CLOUD CURRENTS AND FREE FLOATING PARTICLE CLOUDS (PAS – P. Cl.) AT SPACE THE BETWEEN ATOMS (ARE PAS- PCS).

INDIGENOUS INTERNAL ATOM – PCS

THE INDIGENOUS INTERNAL ELECTRON - PCS, THE INDIGENOUS INTERNAL NEUTRON - PROTON PARTICLE CLOUD CIRCULATION SYSTEMS, THE INDIGENOUS INTERNAL ATOM PARTICLE CLOUD CIRCULATION SYSTEMS. ALSO THEIR RELATED DIFFERENT

PARTICLE INTELLIGENCE SYSTEM CENTERS (PIS) SUCH AS: INDIGENOUS ELECTRON – PIS, INDIGENOUS NEUTRON - PROTON - PIS, AND INDIGENOUS ATOM - PARTICLE INTELLIGENCE SYSTEM CENTERS (I – PIS), AND INDIGENOUS FUNDAMENTAL PARTICLE CIRCULATION SYSTEMS (I. PCS), ALL ARE INSIDE ATOM PCS, AND INSIDE ATOM - PIS CONSTRUCTIONS.

THE NORMAL CNS CONSTRUCTION ELECTRONS, NUCLEONS AND NORMAL ATOM - PIS PRODUCE AND EMIT NORMAL PARTICLE CLOUDS, THESE PARTICLE CLOUDS THROUGH NORMAL PCS TRANSMIT TO DIFFERENT BODY ORGANS AND SYSTEMS, CAUSE NORMAL FUNCTION BODY ORGANS AND BODY SYSTEMS. THE NORMAL FUNCTIONS ORGANS AND BODY SYSTEMS PRODUCE NORMAL BODY SIGNS AND SYMPTOMS.

IN CONTRAST THE ABNORMAL FUNCTIONING AND ABNORMAL CONSTRUCTION CNS ELECTRONS, NUCLEONS, AND ABNORMAL FUNCTIONING CNS- PIS FROM DIFFERENT LOCATION CNS CENTERS, WILL PRODUCE AND EMIT ABNORMAL CNS CENTER RELATED PARTICLE CLOUDS INTO GENERAL BODY PCS.

THESE ABNORMALLY CONSTRUCTED CNS- PARTICLE CLOUDS (S. Y. T. E. – FP – I.I. – P. Cl.) THROUGH BODY'S FUNDAMENTAL PARTICLE CLOUD CIRCULATION SYSTEMS TRAVEL TO PERIPHERALLY RELATED DIFFERENT BODY ORGANS AND BODY SYSTEMS. THE RECIPIENT BODY ORGANS AND BODY SYSTEMS UNDER HIERARCHY PARTICLE CLOUD OPERATION SYSTEMS OF THE DOMINANT CNS – ELECTRONS AND NUCLEONS PARTICLE CLOUD ORDERS, WILL FUNCTION EXACTLY ACCORDING THE PARTICLE CLOUD ORDERS OF DOMINANT CNS ATOM CENTERS.

CONSEQUENTLY, THE ABNORMAL PARTICLE CLOUD ORDER RECIPIENT ORGANS AND SYSTEMS WILL FUNCTION ABNORMAL. AND WILL RESULT PRODUCTION OF ABNORMAL BODY ORGANS, AND BODY SYSTEMS ABNORMAL SIGNS AND SYMPTOMS.

FOR EXAMPLE, THE PRODUCED AND EMITTED ABNORMAL PARTICLE CLOUDS ORDERS (ABNORMAL S. Y. E. T. FP – I.I. – P. Cl.) FROM ABNORMALLY CONSTRUCTED CNS – MUSCULOSKELETAL CENTERS ELECTRONS AND NUCLEONS, MAY TRAVEL TO A GIVEN SPECIFIC NORMAL CONSTRUCTION LIMBS, AND CAUSE FUNCTIONAL PARALYSIS IN NORMALLY CONSTRUCTED BODY ORGAN (NEUROSIS, EPILEPSY).

IN ANOTHER EXAMPLE, WHEN THE PRODUCED AND EMITTED ABNORMAL PARTICLE CLOUDS (S. Y. T. E. – FP. – I.I. – P. Cl.) FROM ABNORMAL CONSTRUCTION GIS – CNS ELECTRONS AND NUCLEONS ENTER INTO PCS. THESE ABNORMAL PARTICLE CLOUDS THROUGH PCS TRAVEL TO RELATED GIS, INTESTINE, STOMACH, ETC. THE ABNORMAL PARTICLE CLOUD RECIPIENT GIS ORGANS (INTESTINE, STOMACH) UNDER ABNORMAL PARTICLE CLOUD ORDERS HAS TO FUNCTION EXACTLY ABNORMAL AS ABNORMAL CONSTRUCTED GIS – CNS ATOM'S PARTICLE CLOUD ORDERS.

THE RESULTS ARE PRODUCTION OF ABNORMAL GASTRO INTESTINAL FUNCTIONS AND PRODUCTION OF RELATED FUNCTION ABNORMALITY SIGNS AND SYMPTOM. (THE DISEASES SIMILAR TO IRRITABLE BOWL SYNDROME, OR GASTRIC RELATED OTHER DISORDERS).

THESE ARE THE EXAMPLES OF PARTICLE CLOUD PRODUCED DISEASES, ALSO THESE ARE THE ATOM SICKNESSES PRODUCED BECAUSE OF MALCONSTRUCTIONS OF CNS ELECTRON, NUCLEON PARTICLE CLOUD COMPOUND CONSTRUCTION ABNORMALITIES. ADDITIONALLY, THESE ARE THE DISEASES WHICH HAVE BEEN PRODUCED BECAUSE OF THE ABNORMAL PARTICLE CLOUD PRODUCTIONS. ALSO SOME OF THESE PARTICLE DISEASES ARE

PRODUCED BECAUSE OF THE ABNORMALITIES IN PCS MALFUNCTIONING, OR BECAUSE OF THE PCS - DISRUPTIONS.

IN ANOTHER EXAMPLE, THE ABNORMALLY PRODUCED PARTICLE CLOUDS (S. Y. T. E. –FP – I.I. – P. Cl.), WHICH HAS BEEN EMITTED FROM SICK CNS – ATOMS OF RESPIRATORY CENTER CNS, THESE PARTICLE CLOUDS HAS BEEN EMITTED FROM ABNORMALLY CONSTRUCTED CNS ELECTRONS, NUCLEONS PARTICLE CLOUD COMPOUND CONSTRUCTIONS. THESE ABNORMAL PARTICLE CLOUDS TRAVEL VIA PARTICLE CIRCULATION SYSTEMS TO THORACO- PULMONARY PIS CENTERS. THESE ABNORMALLY CONSTRUCTED PARTICLE CLOUDS ORDERS AT CHEST – PIS, PRODUCE ABNORMAL THORACO PULMONARY MALFUNCTIONS ACCORDING THE ABNORMAL CNS PARTICLE CLOUD HIERARCHY ORDERS.

CONSEQUENTLY, PRODUCE THORACO PULMONARY MALFUNCTIONS UNDER CLOUDS ORDERS, THE PRODUCED THORACO PULMONARY MALFUNCTIONS CAUSE PRODUCTION OF RELATED ABNORMALITIES IN SIGNS AND SYMPTOMS. (SUCH AS DYSPNEA, COUGH, ETC.) SIMILAR TO ASTHMA AND OTHER DISEASES.

FOLLOWING ARE EXAMPLES FROM I – PCS:

1 – INTERNAL ATOM PCS. AND ATOM'S - PIS.

2 – INTERNAL PERIPHERON – PCS, AND PERIPHERON'S - PIS.

3 - INTERNAL ATOM NUCLEUS – PCS, AND NUCLEI - PIS.

4 – INTERNAL ELECTRON – PCS, AND ELECTRON'S -PIS.

5 - INTERNAL PROTON – PCS, AND PROTON'S - PIS.

6- INTERNAL NEUTRON PCS, AND NEUTRON'S - PIS.

EXAMPLES FROM CELL'S – PCS & PIS

1 - THE INTERNAL CELL PARTICLE CLOUD CIRCULATION SYSTEMS (CELL – PCS), AND CELL'S – PIS CENTERS.

2 - INTERNAL NUCLEI – PCS, AND CELL NUCLEI – PIS CENTERS.

3 - INTERNAL CYTOPLASMIC – PCS, AND PIS OF CYTOPLASM CENTER.

MALFUNCTION IN ANY OF ABOVE MENTIONED CELL PARTICLE INTELLIGENCE SYSTEM CENTERS, OR AT ANY OF THE PARTICLE CIRCULATION SYSTEMS, WILL CAUSE CELL FUNCTION DEFICIENCIES AND FAILURE, AND MALFUNCTIONING WILL PRODUCE RELATED ABNORMAL SIGNS AND SYMPTOMS.

PCS INSIDE LIVING THING'S BODY

COMMUNICATION OF TOTAL BODY CELL POPULATION WITH EACH OTHER, THROUGH PIS, PCS

COMMUNICATION OF TOTAL BODY ELECTRON, NEUTRON, PROTON, ATOM POPULATION BY PCS, WITH EACH OTHER

THE LIVING THINGS OF EARTH HAVE TWO DISTINGUISHED PARTICLE CIRCULATION SYSTEMS (PCS):

1 –THE VERTICAL PCS: THE CNS ELECTRONS, NEUTRONS, PROTONDS PRODUCE AND EMIT FUNDAMENTAL PARTICLE INFORMATION AND IMAGE PARTICLE CLOUDS FROM CNS VOLUNTARY OR INVOLUNTARY DOMINANT CELL CENTER ATOMS

(S. Y. E. T. –FP – I.I. – P. Cl.), UNDER A. S. I. FP. Mol. CIC. THESE PARTICLE CLOUD CURRENTS MOSTLY FLOW VERTICALLY THROUGH TWO BIDIRECTIONAL PARALLEL EFFERENT AND AFFERENT PARTICLE CIRCULATION SYSTEMS (PCS). BETWEEN DOMINANT CNS ELECTRON, NUCLEON CENTERS AND RECIPIENT DIFFERENT BODY ORGSN AND BODY SYSTEM PARTICLE INTELLIGENCE SYSTEM CENTERS. THE ORGAN – PIS, AND BODY SYSTEM – PIS ALL HAVE TO WORK ACCORDING ORDERS OF THESE CNS ORDER PARTICLE CLOUDS.

THE PERIPHERAL BODY SYSTEMS AND BODY ORGANS TOTAL ELECTRON, NEUTRON, PROTON, ATOM POPULATIONS, ALSO THE BODY SYSTEMS AND BODY ORGANS TOTAL CELL NUMBER POPULATIONS ALL HAVE TO FUNCTION CHEMICALLY, PHYSICALLY, BIOLOGICALLY UNDER THESE CNS EMITTED NORMAL OR ABNORMAL PARTICLE CLOUD ORDERS.

2 – THE OBLIQUE PCS (TRANSVERSE –PCS) CURRENT RUN BETWEEN DIFFERENT BODY SYSTEMS AND BODY ORGANS PARTICLE INTELLIGENCE SYSTEM CENTERS (O – PIS, S – PIS): THE DIFFERENT BODY ORGANS AND BODY SYSTEMS PARTICLE INTELLIGENCE SYSTEM CENTERS PRODUCE AND EMIT INFORMATION AND IMAGE ORDER PARTICLE CLOUDS FROM SELF INTO PCS.

THE ORGAN AND BODY SYSTEM EMITTED PARTICLE CLOUDS FROM O – PIS, S – PIS, THROUGH OBLIQUE OR TRANSVERCE PCS TRANSIT FROM ONE ORGAN, OR ONE BODY SYSTEM, TO OTHER ORGANS, OR BODY SYSTEM PARTICLE INTELLIGENCE CENTERS. THESE EMITTED PARTICLE CLOUDS ORDERS AND INFORMATIONS – IMAGE PARTICLE CLOUDS FROM DIFFERENT O- PIS AND S – PIS TRAVEL FROM ONE O – PIS TO OTHER O- PIS, UNDER A. S. I. FP. Mol. CIC. INTERACT WITH EACH OTHER. AND COORDINATE THE DIFFERENT O – PIS AND S- PIS FUNCTIONS WITH EACH OTHER.

THE PRODUCED AND EMITTED NORMAL OR ABNORMAL PARTICLE CLOUDS (S. Y. E. T. – FP – I.I. – P. Cl.) FROM ONE GIVEN O – PIS, OR FROM ONE GIVEN S – PIS INTO OBLIQUE PCS, TRAVEL FROM ONE ORGAN TO OTHER, OR FROM ONE BODY SYSTEM TO THE OTHER, BETWEEN THE DIFFERENT BODY ORGAN'S AND BODY SYSTEM'S PIS.

THE DIFFERENT ORGAN PIS THROUGH THESE OBLIQUE PARTICLE CLOUD INFORMATIONS, REGULATE AND COORDINATE THEIR FUNCTIONS WITH OTHER ORGANS. IN GENERAL, THE ABNORMAL O – PIS, AND ABNORMAL S – PIS PRODUCE AND EMIT ABNORMAL PARTICLE CLOUDS FROM SELF.

THESE ABNORMALLY PRODUCED AND EMITTED PARTICLE CLOUD FROM ONE GIVEN ORGAN'S PIS, WHEN TRAVEL THROUGH OBLIQUE PCS TO OTHER ORGANS PIS, THESE ABNORMAL PARTICLE CLOUDS WILL PRODUCE ABNORMAL BIOLOGICAL FUNCTIONS IN ANOTHER RECIPIENT BODY ORGAN'S PIS.

FOR EXAMPLE, THE ABNORMAL FUNCTIONING DISTAL COLON, RECTAL PARTICLE INTELLIGENCE SYSTEM CENTER MAY PRODUCING AND EMITTING ABNORMAL PARTICLE CLOUDS OF FREQUENCY, DYSFUNCTION, URGENCY, OBSTRUCTION, MALFUNCTION, BLOCKAGE, ETC. THESE ABNORMALLY PRODUCED DISTAL GIS RECTUM –PIS PARTICLE CLOUD ORDERS, TRAVEL AND TRANSIT VIA OBLIQUE - PCS TO ADJACENT DISTAL GUS - PIS CENTERS, UNDER ABNORMALLY ISSUED PARTICLE CLOUDS CAUSING BY MISTAKE URGENCY, FREQUENCY, NONCONTROL, DSYFUNCTIONAL GUS SIGNS AND SYMPTOMS ABNORMALITIES.

THE DISTAL COLON PIS PRODUCED ABNORMAL FUNDAMENTAL PARTICLE INFORMATION IMAGE PARTICLE CLOUDS (FP – I.I. – P. Cl.) IN REGARD FOR HAVING FUNCTION ABNORMALITIES OF FREQUENCY, OBSTRUCTION, URGENCY, BLOCKAGE VIA OBLIQUE PCS TRANSMISSION TO DISTAL GUS - PIS, CAN CAUSE AND

PRODUCE GUS ABNORMALITY IN THE FORMS OF URGENCY, FREQUENCY, PAIN, ETC.

THE ABNORMALITIES IN PIS CONSTRUCTION, PCS ABNORMALITIES, THE ABNORMALITIES IN PARTICLE CLOUD PRODUCTIONS AND PARTICLE CLOUD TRANSMISSION ABNORMALITIES ALL CAN CAUSE IN OTHER ADJACENT ORGANS FUNCTION ABNORMALITIES, AND FUNCTION ABNORMALITY ALSO CAN PRODUCE SIGNS AND SYMPTOMS ABNORMALITIES AS RESULTS.

BODY ORGAN'S PCS & PIS:

ORGANS PARTICLE CLOUD CIRCULATION SYSTEMS (O-PCS), ORGANS PARTICLE INTELLIGENCE SYSTEM (O-PIS)

THE PARTICLE INTELLIGENCE SYSTEM CENTERS (O – PIS) OF EACH GIVEN ORGAN, COMMUNICATE AND CONTROL (THROUGH PARTICLE CLOUD ORDER SYSTEMS) THE ORGAN'S TOTAL ELECTRON, NEUTRON, PROTONS POPULATION'S CHEMICAL, BIOLOGICAL, AND PHYSICAL FUNCTIONS.

THE PRODUCED PARTICLE CLOUD ORDERS FROM O – PIS, TRAVEL VIA ORGAN'S –PCS, TO ORGAN'S ELECTRON, NUCLEON, ATOM, CELL TOTAL POPULATIONS. AND ALL BODY ORGAN'S CELLS AND ATOMS MUST WORK AND FUNCTIONS UNDER PARTICLE CLOUD ORDERS WHICH HAS BEEN PRODUCED AND EMITTED BY O – PIS.

AT SAME TIME ALL O – PIS HAS TO FUNCTION ACCORDING THE INCOMING PARTICLE CLOUD ORDERS FROM DOMINANT CNS – ELECTRONS, NUCLEONS, ATOMS AND CELL'S PARTICLE CLOUD ORDERS, THE CNS AND S – PIS ATOM'S PRODUCE AND EMIT

ORGAN FUNDAMENTAL PARTICLE CLOUD ORDERS AND CONTROL ALL BODY ORGANS PIS CENTERS.

Abnormal Particle Cloud Compound Construction of CNS
Electrons, Protons, Neutrons

MALFUNCTION OF PIS

MALFUNCTION OF PCS

Atom Sicknesses, PCS diseases, Particle Cloud Disorders

THE ABNORMAL CNS ELECTRON NUCLEON CONSTRUCTIONS WITH ABNORMAL PARTICLE CLOUD COMPOUNDS, WILL CAUSE PRODUCTION AND EMISSION OF ABNORMAL PARTICLE CLOUDS (ATOM SICKNESSES) FROM CNS ATOMS, TRANSMISSION OF THESE PARTICLE CLOUDS TO RECIPIENT ORGANS WILL PRODUCE MALFUNCTION IN RECIPIENT ORGAN'S BIOLOGICAL FUNCTIONS, AND WILL PRODUCE ABNORMAL SYMPTOMS AND SIGNS. THE DYSFUNCTION OF THE RECIPIENT BODY SYSTEM'S PARTICLE INTELLIGENCE SYSTEM CENTER ALSO IS THE CAUSES OF FUNCTIONAL ORGAN DISORDERS. ABNORMALITIES AND DISRUPTION OF PARTICLE CLOUD CIRCULATION SYSTEMS ALSO ARE CAUSE FOR LARGE NUMBERS OF ORGAN FUNCTIONAL DISORDERS.

FOR EXAMPLE, DEFECTIVE CNS – ELECTRON NUCLEON CONSTRUCTIONS WITH DEFECTIVE FALSE INFORMATION AND IMAGE PARTICLE CLOUDS COMPOUNDS CONSTRUCTIONS OF CNS

- GIS AND CNS - GUS CENTERS (ASUME THE GIS AND GUS ELECTRONS AND NUCLEONS CONSTRUCTED WITH SICK URGENCY, FREQUENCY, PAIN, PARTICLE CLOUD COMPOUND CONSTRUCTIONS, EXIST AT DISTAL GUS, OR IN DISTAL GIS). EMISSION OF THESE ABNORMAL SIGNS AND ABNORMAL SYMPTOMS INFORMATIONS AND IMAGE PARTICLE CLOUDS FROM CNS – PIS CENTERS, AND TRANSMISSION OF THESE PARTICLE CLOUDS TO RECIPIENT GIS, OR GUS PARTICLE INTELLIGENCE SYSTEM CENTERS. WILL CAUSE THE RECIPIENT GIS, AND GUS ORGANS PIS AT DISTAL COLON, AND DISTAL GUS FUNCTION EXACTLY ACCORDING THE ABNORMAL PARTICLE CLOUD ORDERS WHICH ISSUED BY SICK ATOM CONSTRUCTION CNS – PIS ELECTRONS AND NUCLEONS. CNS ORDERED FALSE CONSTRUCTION PARTICLE CLOUDS CAUSING PRODUCTION OF

FALSE FREQUENCY, URGENCY, PAIN, BLOCKAGE AT DISTAL GIS, AND GUS UNDER PARTICLE CLOUD ORDERS. WHICH CAUSED THE DISTAL GUS, AND DISTAL GIS SYSTEMS PARTICLE INTELLIGENCE SYSTEM CENTERS FUNCTION ABNORMAL ACCORDING CNS ATOM ORDERS.

THE MALFUNCTIONING DISTAL GIS, OR DISTAL GUS PARTICLE INTELLIGENCE SYSTEM CENTERS CAN PRODUCE SECONDARY ABNORMAL PARTICLE CLOUD ORDER PRODUCTIONS, AND TRANSMISSION OF THOSE SECONDARILY ABNORMAL PARTICLE CLOUDS THROUGH OBLIQUES – PCS TO ADJACENT OR FAR AWAY ORGANS VIS PCS, CAN INFLICT MULTI ORGANS PIS DISEASES UNDER DOMINOS EFFECTS.

COLD FUNDAMENTAL PARTICLE CLASSES
(Cold Particle related Diseases)
ANGINA CRANIUM SYNDROM

MOSTLY IN WINTER COLD, THE EXOGENOUS COLD FUNDAMENTAL PARTICLES DIRECTLY AIRBORNE FROM SKIN TRANSIT INTO CRANIUM'S (SKULLS) INTERNAL ELECTRONS AND NUCLEONS PARTICLE MOLECULAR CONSTRUCTIONS, INTERACT WITH MOLECULAR CONSTRUCTIONS OF SKULL'S INTERNAL ELECTRONS AND NUCLEONS PARTICLE COMPOUNDS AND PRODUCE COLD PARTICLE COMPOUNDS (COLD FUNDAMENTAL PARTICLE COMPOUND NEO GENESIS).

THESE NEO MOLECULAR CONSTRUCTIONS SIGNS ARE PRODUCTION OF FREEZINF COLD CRANIUM WITH EXCRUCIATING SEVERE NON STOP ANGINAL CRANIAL PAINS. WHICH NOT RESPOND TO ORDINARY PAIN MEDICINES, MOST OF THESE AUTONOMOUS CHEMICAL INTERACTIONS ARE REVERSIBLE FOLLOWING REWARMING AND PROVIDING THERMAL FUNDAMENTAL PARTICLES THE COLD PARTICLES BREAK DOWN AND LEAVE THE AREA, NEO -THERMAL PARTICLE COMPOUNDS CONSTRUCT IN THEIR PLACE, ANGINAL PAIN DISAPPEAR. THE THERAPY IS IMMEDIATE TOPICAL APPLICATION OF THERMAL FUNDAMENTAL PARTICLES.

TYMPANIC ANGINA

THE COLD FUNDAMENTAL PARTICLES IN WINTER AT ELDERLY AND COMPROMISED IMMUNITY CASES AIRBORNE TRAVEL INTO EXTERNAL EAR CANAL, DIRECTLY TRANSIT FROM SKIN INTO EXTERNAL EAR CANALS AND TYMPANIC MEMBRANE'S INTERNAL ELECTRONS AND NUCLEON SUBSYSTEM UNITS, INTERACT WITH INTERNAL ATOM PARTICLE COMPOUND CONSTRUCTIONS, COLD PARTICLES CAUSE MOSTLY SLOW DOWN AND STAGNATION OF BIOLOGICAL AND CHEMICAL INTERACTIONS, DECREASE IN

CIRCULATION SYSTEMS, VASOCONSTRICTION, TISSUE ANOXIA IN ELDERLY WHICH THE SIGNS ARE EXCRUCIATING ANGINAL TYPES PAINS OF TYMPANIC MEMBRANES. SEVERE ANGINAL PAIN NOT RESPOND TO ORDINARY PAIN MEDICATIONS. THE EXAMINATION AND PARA CLINICAL TESTS ARE NORMAL. TREATMENT FOR ANGINAL TM IS SIMILAR TO ANGINA CRANIUM, PROVIDING THERMAL PARTICLES REVERSE PROCESS IN SHORT TIME. NON TREATED CASES CAUSE SECONDARY INFLAMMATIONS AND INFECTION COMPLICATIONS.

ANGINA PEDIS

(COLD FUNDAMENTAL PARTICLES INJURIES AT OTHER ORGANS)

THE COLD FUNDAMENTAL PARTICLES ALSO INVADE THE OTHER DIFFERENT BODY TISSUES AND CELLS INTERNAL ELECTRON AND NUCLEON CONSTRUCTIONS. IN ELDERLY THE COLD FUNDAMENTAL PARTICLES COMMONLY ENTER THROUGH CUTANEOUS TISSUES OR OTHER ROUTES INTO FEET'S INTERNAL ATOM PARTICLE MOLECULAR STRUCTURES. PRODUCE ISCHEMIA, VASOCONSTRICTION, ANGINAL ISCHEMIC SEVERE NON STOPPING PAINS.

AS A GENERAL RULE, THE INTERNAL ATOM DISEASES AND INJURIES MOSTLY PRODUCE THE SAME OR SIMILAR KINDS SIGNS AND SYMPTOMS AS EXACTLY SIMILAR TO TISSUE AND CELL DISEASES SIGNS AND SYMPTOMS. TREATMENT IS REWARMING AND RELIEF OF PAINS. **SICK ATOM'S SIGNS -SYMTOMS ARE SIMILAR TO SICK TISSUE SIGNS AND SYMPTOMS.**

PSYCHE GENESIS
Thought Current Genesis in Living Things CNS Electrons, Nucleons
PHENOMENON OF S. Y. E. T.– F.P. – I.I. – P.cl. STORAGE INSIDE CNS ATOMS (STORAGE OF INTELLIGENCE)

AIRBORNE EXOGENOUS EMITTED BIOFRIENDLY ELECTRIC FUNDAMENTAL PARTICLE CLOUDS, LIGHT, SONIC AND THERMAL FUNDAMENTAL PARTICLE INFORMATION IMAGE PARTICLE CLOUDS (S. E. T. Y. F.P. – I.I.- P. cl.) FROM DIFFERENT ENVIRONMENTAL SUBJECTS, TRAVEL VIA AIR TO DIFFERENT LIVING THINGS SENSORY ORGANS. THE RECIPIENT LIVING THINGS SENSORY ORGAN ATOM'S ATTRACTIVE GRAVITON FORCES ATTRACT AND CAPTURE THESE INCOMING PARTICLE CLOUDS, THEREAFTER TRANSIT INTO DIFFERENT CNS RELATED BRAIN CENTERS. SONIC PARTICLE CLOUDS MOSTLY TRAVEL TO AUDITARY CNS HEARING CENTER. AND LIGHT PARTICLE CLOUDS TRANSIT INTO VISION CNS OF BRAIN.

THE INCOMING EXOGENOUS S. Y. T. E. –FP – I.I. – P. Cl. ENTER INTO CNS INTERNAL ELECTRON NUCLEON SUBSYSTEM UNITS CHEMICAL LAB.S, UNDER A. S. I. FP. Mol. CIC. COMBINE WITH INDIGENOUS PRE-EXISTING INTERNAL ELECTRON, NUCLEON PARTICLE COMPOUNDS, STORE OUTSIDE WORLD INFORMATIONS IN PARTICLE CLOUD KNOWLEDGE AND SCIENCE IN PARTICLE CLOUD FORMS INSIDE CNS ATOMS.

THROUGH THE A. S. I. FP. Mol. CIC. EXOGENOUS INFORMATION AND IMAGE PARTICLE CLOUDS, WHICH ARE OUTSIDE WORLD SUBJECTS SCIENCE, KNLOWLEDGE INFORMATION AND IMAGE COPIES IN PARTICLE CLOUD FORMS. THESE INCOMING PARTICLE CLOUDS COMBINE WITH INTERNAL ATOM PARTICLE COMPOUNDS, CONSTRUCT CNS ELECTRON, PROTON, NEUTRON PARTICLE CLOUD COMPOUND CONSTRUCTIONS. AND STORE OUTSIDE WORLD KNOWLEDGE, SCIENCE, INFORMATION AND IMAGE ABOUT DIFFERENT SUBJECTS, INSIDE THE CNS ATOMS.

THIS IS PHENOMENON OF KNOWLEDGE STORAGE SYSTEMS INSIDE CNS ATOMS. THIS IS PHENOMENON OF NEO- S. Y. – F.P. –I.I. – P.cl. - COMPOUND CONSTRUCTION (NEO-GENESIS OF PARTICLE –CLOUD COMPOUNDS) INSIDE CNS INTERNAL ELECTRON NUCLEONS. AS WELL AS THIS IS PHENOMENON OF STORING SONIC – PARTICLE INFORMATION – IMAGE PARTICLE CLOUDS, AND LIGHT FUNDAMENTAL PARTICLE INFORMATION IMAGE PARTICLE CLOUDS INSIDE THE CNS ELECTRONS AND NUCLEONS, IN THE PARTICLE INFORMATION IMAGE PARTICLE CLOUD CONPOUND FORMS. CONSTRUCT CNS ATOM STRUCTURES WITH LIGHT, SONIC, ELECTRIC, THERMAL PARTICLE CLOUD –COMPOUND CONSTRUCTIONS (S. Y. E. T.– F.P. – I.I.- P. Cl. - COMP.).

 DURING LIFE, ANIMALS COMPILE STORAGE OF DIFFERENT ENVIRONMENTAL SUBJECTS PARTICLE CLOUD INFORMATIONS AND IMAGES (S. Y. T. E. – F.P. – I.I. – P. Cl.) IN PARTICLE COMPOUND FORMS, STORED INSIDE CNS ELECTRONS, NUCLEONS. THESE INFORMATIONS AND IMAGES PARTICLE COMPOUND STORAGES INSIDE CNS ELECTRONS, NUCLEONS ARE STORAGE OF KNOWLEDGE.

WHICH THESE STORED KNOWLEDGE AT ALL TIMES ARE AVAILABLE FOR RETRIEVAL, AND RELEASE TO OUT OF CNS ELECTRONS NUCLEON CONSTRUCTIONS. THIS IS PHENOMENON OF RELEASE AND RETRIEVAL OF INFROMATION FROM CNS ATOMS TO OUTSIDE WORLD.

THESE STORED PARTILE CLOUD COMPOUNDS ARE BUILDING BLOCKS OF KNOWLEDGE, SCIENCE, INTELLIGENCE CONSTRUCTIONS, AND ARE USED IN LEARNING, EDUCATIONS, TRAININGS, AND INTELLIGENCE CONSTRUCTIONS INSIDE CNS ELECTRONS NUCLEONS OF LIVING THINGS, AS WELL AS NON-LIVE ELECTRONS-NUCLEONS PARTICLE COMPOUND CONSTRUCTIONS. THESE STORED INFORMATIONS AND IMAGES ARE AVAILABLE FOR RETRIEVAL TO OUTSIDE, AS FREE CLOUDS AIRBORNE FORMS AGAIN.

THIS IS PHENOMENON OF PSYCHE-GENESIS AND STORAGE OF SCIENCE AND INFORMATION IMAGE INSIDE ELECTRONS NUCLEONS IN PARTICLE COMPOUND FORMS, INTERACTIONS OF THESE SONIC LIGHT FUNDAMENTAL PARTICLE COMPOUND INFORMATION IMAGE PARTICLE CLOUDS WITH EACH OTHER AS WELL AS WITH EXOGENOUS INCOMING S. Y. – F.P. – I.I. – P. Cl. PRODUCE THOUGHT CURRENTS, PSYCHE.

THE THOUGHT CURRENTS ARE SONIC, LIGHT, ELECTRIC FUNDAMENTAL PARTICLE INFORMATION IMAGE PARTICLE CLOUD CURRENTS AND INTERACTIONS OF THESE PARTICLE CLOUD CURRENTS WITH EACH OTHER INSIDE CNS ELECTRONS NUCLEONS PRODUCE PSYCHE AND THOUGHT SYSTEMS.

Intelligence Genesis and accumulation of Knowledge inside CNS Electrons and Nucleons

ATTRACTION OF S.Y. –F.P. – I.I. – P. Cl., AND PARTICLE CLOUDS FROM AIR, BY SENSORY ORGANS

PHENOMENON OF PSYCHE GENESIS AND THOUGHT CURRENT GENESIS

MEMORIZATION PHENOMENON

THE ATTRACTIVE GRAVITON FORCES OF SENSORY ORGANS ELECTRONS AND NUCLEONS, ATTRACT THE INCOMING EX.- S. – F.P. – I.I. – P. Cl., EX. - Y. - F.P.- I.I. – P. Cl., EX. - T. – F.P. – I.I. – P. Cl., EX. - E. –FP – I.I. – P. Cl., COLD PARTICLES, ETC., FROM EX. PARTICLE CLOUD CIRCULATION SYSTEMS OF SPACE (AIR).

THE ENVIRONMENTAL SUBJECTS PRODUCE AND EMIT EXACT COPIES OF SELF, WHICH THOSE SUBJECT COPIES ARE MADE FROM DIFFERENT KIND SONIC, LIGHT, ELECTRIC, AND THERMAL FUNDAMENTAL PARTICLE INFORMATIONS AND IMAGE PARTICLE CLOUDS (FP – I.I. – P. Cl.). THESE

INFORMATION AND IMAGE PARTICLE CLOUDS ARE FUNDAMENTAL PARTICLE MADE COPIES OF EXISTING THINGS OF OUTSIDE WORLD.

THESE INCOMING EXOGENOUS DIFFERENT KIND INFORMATION AND IMAGE PARTICLE CLOUDS ENTER INTO SENSORY ORGANS INTERNAL ELECTRON, NUCLEON SUBSYSTEM UNITS CHEMICAL LAB.S, THESE INCOMING EX. PARTICLE CLOUDS TRANSIT (VIA INDIGENOUS PARTICLE CLOUD CIRCULATION SYSTEM (I – PCS). FROM SENSORY ORGAN ATOM CENTERS INTO DIFFERENT PARTICLE CLOUD RELATED CNS CENTERS. WHICH THOSE RECIPIENT CNS CENTERS ATOMS ARE IN NEEDS OF THESE INCOMING EX. FP - I.I. – P. Cl. IN ORDER TO USE THESE NEEDED PARTICLE CLOUDS INSIDE THEIR CHEMICAL LAB.S, AND UNDER REG. A. S. I. FP. Mol. CIC. CONSTRUCT THEIR CNS INTERNAL ELECTRON, NUCLEON, NEEDED FP – I.I. – P. Cl. – COMPOUND CONSTRUCTIONS. (NEO GENESIS OF FUNDAMENTAL PARTICLE COMPOUND CONSTRUCTIONS OF CNS ATOMS, WITH USE OF EX. FP – I.I. – P. Cl. IS PHENOMENON OF STORAGE OF INFORMATION CLOUDS).

THE AUDITARY CNS CENTERS ELECTRON, NUCLEON FOR THEIR ELECTRON NUCLEON NEO GENESIS NEED SONIC PARTICLE CLOUDS TO BUILD THEIR INTERNAL ATOM SONIC PARTICLE CLOUD COMPOUND CONSTRUCTIONS (NEO GENESIS OF CNS PARTICLE COMPOUND CONSTRUCTIONS WITH SONIC PARTICLE CLOUDS).

THE SONIC FUNDAMENTAL PARTICLE CLOUD SENSING AND CAPTURING SENSORY ORGANS, SUCH AS EAR BIOLOGICAL DEVICE, UNDER ORGAN ATOM'S STRONG ATTRACTIVE

GRAVITON FORCES TO ATTRACT SONIC PARTICLES, THESE ORGAN'S ATOMS ATTRACT AND CAPTURE SONIC CLOUDS PARTICLES FROM AIR, THEREAFTER TRANSIT THE CAPTURED S. FP. – I.I. – P. Cl. FROM EAR SENSARY ORGAN, INTO AUDITARY CNS CENTERS INTERNAL ELECTRON NUCLEONS PARTICLE SUBSYSTEM UNITS CENTERS.

IN SONIC CNS THE PARTICLE CLOUDS COMBINE WITH INTERTNAL ELECTRON NUCLEON PARTICLE COMPOUNDS, BECOME PART OF AUDITARY CNS CENTERS ELECTRON NUCLEON PARTICLE COMPOUND MOLECULAR CONSTRUCTIONS. AND STORE S –FP –I.I. –P. Cl. COMPOUNDS INSIDE CNS ELECTRONS, NUCLEONS (INTELLIGENCE GENESIS PHENOMENON BY SONIC INFORMATION CLOUDS).

THE EYES LIGHT SENSING SENSORY ATOMS, WHICH HAS STONG ATTRACTIVE GRAVITON FORCES TO ATTRACT AND CAPTURE INCOMING AIRBORNE EXOGENOUS (EX - Y – FP – I.I. – P. Cl.) LIGHT PARTICLE CLOUDS FROM AIR. THEREAFTER THE CAPTURED LIGHT PARTICLE CLOUDS TRANSIT FROM LIGHT SENSING BIOLOGICAL DEVICE OF EYE, VIA INDIGENOUS PCS, INTO CNS VISION CENTER'S INTERNAL ELECTRON, NUCLEON SUBSYSTEM UNITS CHEMICAL LAB.S.

THE Y. FP - I.I. – P. Cl. INSIDE CNS VISION CENTER ELECTRONS, NUCLEONS CHEMICAL LAB.S, UNDER A. S. I. FP. MOL. CIC. COMBINE WITH INTERNAL ELECTRON AND NUCLEON PARTICLE COMPOUNDS, AND PRODUCE VISION CENTER'S CNS ELECTRON NUCLEON CONSTRUCTIONS, WHEN USING EX. Y – FP – I.I. – P. Cl. – COMPOUND CONSTRUCTIONS. THIS IS STORAGE OF Y. FP – I.I. – P. Cl.

INFORMATION AND KNOWLEDGE INSIDE THE CSN ATOMS, IN VISION CENTER.

THIS IS PHENOMENON OF INTELLIGENCE GENESIS AND KNOWLEDGE ACCUMULATION INSIDE CNS ELECTRONS AND NUCLEONS, THE KNOWLEDGE AND INTELLIGENCE STORED IN PARTICLE CLOUD COMPOUND FORMS INSIDE THE CNS ELECTRONS, NEUTRONS, PROTONS, AND ATOMS.

THE MAJORITIES OF EARTH ELECTRON, NEUTRON, PROTON PARTICLE CLOUD COMPOUNDS HAVE BEEN CONSTRUCTED FROM PRESENTLY UNKOWN BIOFRIENDLY NONDISCOVERED FUNDAMENTAL PARTICLES (BF- ND – FP). WHICH THESE EXOGENOUS NON DISCOVERED FUNDAMENTAL PARTICLE INFORMATION AND IMAGE PARTICLE CLOUDS, TRAVEL AIRBORNE OR TRANSIT THROUGH PAS – PCS, AND ENTER INTO CNS ELECTRONS, NUCLEONS SUBSYSTEM UNITS CHEMICAL LAB.S, COMBINE WITH INTERNAL ELECTRON NUCLEON SUBSYSTEM UNITS PARTICLE COMPOUNDS AND PRODUCE DIFFERENT CNS CENTERS INTERNAL ELECTRON NUCLEON BF – ND – FP - COMPOUND MOLECULAR CONSTRUCTIONS OF THE CNS ATOMS.

THE OTHER KNOWN FUNDAMENTAL PARTICLES OF THE EARTH SUCH AS BIOFRIENDLY THERMAL PARTICLE CLOUDS, BIOFRIENDLY ELECTRIC FUNDAMENTAL PARTICLE CLOUDS, ALSO VIA AIRBORNE, OR THROUGH PAS - PCS ENTER INTO DIFFERENT CNS CENTER INTERNAL ATOMS SUBSYSTEM UNITS THERMAL CNS CENTERS, OR OTHER PARTICLE RELATED CNS CENTERS, COMBINE WITH CNS INTERNAL ELECTRON, NUCLEON PARTICLE COMPOUNDS, AND STORE OUTSIDE THERMAL, ELECTRIC, SONIC, LIGHT IMAGES AND

INFORMATION KNOWLEDGES AND SCIENCE IN PARTICLE CLOUD COMPOUND FORMS INSIDE THE CNS ATOMS.

THE INTERACTIONS OF DIFFERENT EXOGENOUS PARTICLE CLOUDS, WHEN THEY ARE INTERACTING WITH DIFFERENT INDIGENOUSLY PRODUCED S. Y. T. E. –FP – I.I. – P. Cl., AND CONCURRENTLY BOTH OF THESE CHAINS AND CYCLES INTERACTING WITH DIFFERENT INTERNAL ELECTRON, NUCLEON S. Y. T. E. FP – I.I. – P. Cl. COMPOUND MOLECULES.

ADDITIONALLY, ALL OF ABOVE MILLIONS TO BILLIONS ENORMOUS CONCURRENTLY RUNNING DIFFERENT DEGENERATIVE AND REGENERATIVE A. S. I. FP. Mol. CIC. GOING ON FOR LIFE, DAY AND NIGHT WITH PARTICLE PRECISION ACCURACIES, BETWEEN HUNDREDS TRILLIONS OF DIFFERENT CNS ELECTRONS, NEUTRONS, PROTONS, ALL COORDINATING AND COOPERATING THEIR BIOLOGICAL, CHEMICAL, PHYSICAL FUNCTIONS WITH EXPONENTIAL SPEEDS OF INTERACTIONS, AND NO MISTAKES ALLOWED UNDER THESE SYSTEMS. ABOVE INTERACTIONS OF THE TOTAL CNS ELECTRONS, PROTONS, NEUTRONS FUNCTIONS WITH EACH OTHER BIOLOGICALLY SENSE AS THOUGHT CURRENTS AND PSYCHE.

THIS IS PHENOMENON OF PSYCHE GENESIS AND THOUGHT CURRENT GENESIS, WHICH FUNDAMENTAL PARTICLE INTELLIGENCE SYSTEM CENTERS INVENTED AND CONSTRUCTED THESE SYSTEMS DURING THE MOLECULAR EVOLUTION, UNDER A. S. I. FP. Mol. CIC. FOR THEIR OWN SELF INTEREST AND SURVIVAL.

Particle Cloud Retrieval Phenomenon (Memorization and recalling)

The Large Particle Molecular constructions inside CNS Electrons, Nucleons (such as S. Y. T. E. – FP – I.I. – P. Cl. Compounds), Under Degenerative (Reverse Regenerative) A. S. I. FP. Mol. CIC. break down into small size free Information and Image Particle Cloud Molecules.

This is S. Y. T. E. – FP – I.I. – P. Cl. - Retrieval Phenomenon, the Retrieval process biologically senses as Memorization of freed Particle Clouds (S. Y. E. T. – FP – I.I. – P. Cl.). which these Particle Clouds were stored inside CNS Electrons and Nucleons in Particle Cloud Compound Forms.

THE REVERSE OF REGENERATIVE A. S. I. FP. Mol. CIC. (UNDER DEGENERATIVE A. S. I. F.P. Mol. C.I.C.) CAUSE BREAK DOWN OF LARGE CONSTRUCTED INTERNAL ELECTRON NUCLEON PARTICLE CLOUD COMPOUND CONSTRUCTION MOLECULES INTO SMALLER ORIGINAL PARTICLE CLOUD S. Y. T. E. –FP – I.I. – P. Cl. MOLECULES. THESE FREED INFORMATION AND IMAGE PARTICLE CLOUDS TRAVEL THROUGH CNS – PCS AND BIOLOGICALLY SENSED AS MEMORIZATION THOUGHT CURRENTS. THIS IS RE GENESIS AND RETRIEVAL OF FREE PARTICLE CLOUDS, AND IT

IS MEMORIZATION PHENOMENONS, (THIS IS PART OF RETRIEVAL PROCESS).

Universe under Graviton Laws and Orders

The Graviton Force is Physical Property of Mass

==============================

THE GRAVITON FORCE AND THE CHEMICAL COMBINATIONS

UNDER ATTRACTIVE GRAVITON FORCE TWO COMPATIBLE MASS UNITS ATTRACT EACH OTHER AND COMBINE CHEMICALLY. UNDER REPULSIVE GRAVITON FORCE ORDERS TWO NON –COMPATIBLE MASS REPULSE EACH OTHER AND REFUSE CHEMICAL COMBINATION. GRAVITON FORCE QUANTITY HAVE DIRECT RELATIONSHIP WITH MASS QUANTITY.

IN CHEMICAL COMBINATION BETWEEN THE MULTIPLE MASS UNITS, THE DOMINANT MASS STAY IN NUCLEUS POSITION, AND FROM CENTRAL LOCATION ATTRACT OR REPULSE RECESSIVE MASSES. GRAVITON LAW AND ORDER APPLY FOR ALL MASSES, FROM FUNDAMENTAL PARTICLE TO UNIVERSE SIZE MASS UNITS.

UNDER THE REGENERATIVE AUTONOMOUS SEQUENTIAL CHEMICAL COMBINATIONS, THE ATTRACTIVE GRAVITON

FORCE CAUSE CHEMICAL INTERACTIONS TO TAKE PLACE. BUT UNDER REPULSIVE GRAVITON ORDERS, THE LARGE MOLECULAR CONSTRUCTIONS UNDER DEGENERATIVE AUTONOMOUS SEQUENTIAL CHEMICAL INTERACTIONS CYCLES BREAK DOWN TO SMALL MOLECULES, AND REFUSE TO COMBINE.

THE ABOVE RULE APPLIES TO ALL OTHER AUTONOMOUS SEQUENTIAL CHEMICAL INTERACTION CYCLE BETWEEN ATOM BASE MASS UNITS, NANO UNIT BASE MASS UNITS, CELL BASE LARGE MICRO UNIT MASS SIZES, AS WELL AS THE GARVITON RULES AND LAWS ALSO APPLY TO LARGE MACRO UNIT MASS AUTONOMOUS SEQUENTIAL INTERACTION.

UNDER ATTRACTIVE GRAVITON FORCES, TWO MICRO UNIT MASSES ATTRACT EACH OTHER, TWO CELLS COMBINE AND BECOME ONE CELL. IN ANOTHER SITUATIONS, UNDER REPULSIVE GRAVITON FORCES ONE CELL DEVIDE INTO TWO CELLS, THE PRODUCED TWO CELLS REPULSE AND REFUSE COMBINATION AND SEPARATE.

THE COMBINATION OR REPULSION OF LARGE MASSES SUCH AS PLANETS, GALAXIES, AS WELL AS COMBINATION OR REPULTION OF TWO DIFFERENT UNIVERSE WITH EACH OTHER, ALSO TAKE PLACE UNDER GRAVITON LAWS AND ORDERS.

IN COMBINATION OF TWO UNIVERSAL MASS UNIT WITH EACH OTHER, ALWAYS HEAVY DOMINANT UNIVERSE, WHICH HAVE BEEN CONSTRUCTED FROM DENSE MULTIPLE GALAXY, MULTIPLE BLACK HOLE POPULATION, WITH DOMINANT GRAVITON FORCE, ALWAYS THE DOMINANT UNIVERSE TAKE POSITION IN CENTER, BUILD NUCLEUS

COMPARTMENT OF UNIVERSE IN CENTER (NUCLEUS - UNIVERSE).

THE PERIPHERAL COMPARTMENT UNIVERSE (PERIPHERON) CONTROLLED UNDER ATTRACTIVE NUCLEAR UNIVERSE GRAVITON FORCES. THE PERIPHERON UNIVERSE HAVE FEW STARS, PLANETS SCATTERED FAR FROM EACH OTHER IN VAST SPACE.

THERE ARE MULTIPLE INDEPENDENT DIFFERENT SIZE UNIVERSES ONE AFTER THE OTHER, WHICH ONE INDEPENDENT UNIVERSE BORDERS THE OTHER NEIGHBORING UNIVERSE WITH DISTINCT BORDERS.

EACH INDEPENDENT UNIVERSE HAVE TWO DISTINCT CONSTRUCTION COMPARTMENTS: NUCLEUS OR CENTRAL COMPARTMENT OF UNIVERSE. AND PERIPHERON COMPARTMENT OF UNIVERSE IN PERIPHERY.

RULES, LAWS AND ORDERS OF THE GRAVITON IN UNIVERSE
UNDER MOLECULAR EVOLUTION ORDERS

UNDER GRAVITON FORCES THE NANO UNITS, MICRO UNITS, AND MACRO UNIT MASSES EITHER COMBINE UNDER

ATTRACTIVE GRAVITON FORCES, OR STAY APART BY REPULSIVE GRAVITON ENERGY ORDERS.

THE MORE MASS QUANTITY, THE MORE GRAVITON FORCE QUANTITY. THE MASS WHO POSSESSES MORE GRAVTON, IT IS POWERFUL AND DOMINANT, THROUGH GRAVITON ENERGY FORCES THE MASSES EITHER ATTRACT EACH OTHER AND COMBINE, OR UNDER REPULSIVE GRAVITON ENERGY FORCES MASSES REFUSE COMBINATIONS AND STAY APPART.

ANY GIVEN MASS UNIT ADDITIONAL TO GRAVITON FORCE, POSSESSES OTHER ENERGY CONTENTS AS WELL:

FOR EXAMPLE, A VIOLET FUNDAMENTAL PARTICLE NANO MASS UNIT, ADDITIONAL TO GRAVITON FORCE, POSSESSES TWO ADDITIONAL QUANTUM ELECTRIC ENERGY, AND QUANTUM LIGHT ENERGY IN ANY GIVEN VIOLET FUNDAMENTAL PARTICLE'S CONTENTS (WHICH COLLECTIVELY VIOLET FUNDAMENTAL PARTICLE NANO MASS UNIT ALTOGETHER POSSESSES THREE DIFFERENT KIND QUANTUM ENERGY INHERITENTANCES).

THE LESS WEIGHT MASS UNITS POSSESSES LESS QUANTITY GRAVITON ENERGY FORCES, AND LESSER QUANTITY OTHER ENERGY CONTENTS. IN CONTRARY THE HIGHER WEIGHT MASS UNITS POSSESSES HIGHER QUANTITY GRAVITON ENERGY FORCES, AND HIGHER QUANTITY OTHER KIND ENERGY CONTENTS.

THE HIGHER WEIGHT MASS UNITS DOMINATE AND CONTROL THE LOWER WEIGHT MASS UNITS.

IN CHEMICAL COMBINATION UNDER AUTONOMOUS SEQUENTIAL CHEMICAL INTERACTION CYCLES OF NUMEROUS DIFFERENT KINDS, THE POWERFUL DOMINANT MORE WEIGHT MASS UNITS, WITH MORE GRAVTON FORCES, MORE QUANTITY OTHER ENERGY CONTENTS, MOSTLY TAKE DOMINANT NUCLEUS POSITION IN CENTRAL LOCATION (PRODUCE A NUCLEUS).

THE DOMINANT MASS UNIT FROM NUCLEUS POSITION THROUGH POWERFUL ATTRACTIVE OR REPULSIVE GRAVITON FORCES, EITHER ATTRACT RECESSIVE LESS WEIGHT MASS UNITS, AND CHEMICALLY COMBINE TOGETHER, AAND PRODUCE NEW MOLECULAR COMPOUNDS.

OR THE DOMINANT MASS UNITS FROM NUCLEUS POSITIONS THROUGH POWERFUL REPULSIVE GRAVITON FORCES REPULSE UNWANTED OTHER RECESSIVE MASSES FROM THEIR PERIPHERON, AND REFUSE CHEMICAL COMBINATIONS. ALWAYS THE DOMINANT MASS UNITS STAY IN CENTER AND CONTROL PERIPHERY FROM THERE, AND RECESSIVE MASS UNITS MOSTLY STAY IN PERIPHERON.

ALL LAWS, ORDERS, AND INTERACTIONS OF DIFFERENT MATTERS (FUNDAMENTAL PARTICLES), AND ELEMENTS WITH EACH OTHER ACROSS THE WHOLE UNIVERSE, IN ANY GIVEN LOCATION ALWAYS ARE RELATIVELY CORRECT.

UNIT OF GRAVITON

EXISTING GRAVITON FORCE QUANTITY IN ONE LIGHT FUNDAMENTAL PARTICLE (AIRBORNE),

IS EQUAL TO ONE QUANTUM UNIT GRAVITON FORCE IN PLANET EARTH LOCATION.

GRAVITON IS ENERGY IN MASS CONTENT, WHICH DIFFERENT MASS UNITS USE THIS FORCE TO CONTROL OTHER MASS UNITS, OR TO BE CONTROLLED BY OTHER MASS UNITS. DIFFERENT MASS UNITS POSSESSES DIFFERENT QUANTITIES GRAVITON FORCES. GRAVITON IS MATTER'S INHERITANCE PROPERTIES.

Genesis of Universe under Graviton force and Molecular Evolution

Existing Mass Unit's - PIS used Graviton Forces
and Cloud Orders constructed Existing Things of Universe

under "Nano - Micro - Macro Unit" PIS emitted Clouds, Graviton constructed Earth's existing Things.

THE POST GALACTICAL AND POST UNIVERSAL EXPLOSION DUSTS COMPOSED FROM DIFFERENT KINDS MASS UNITS OF THE FUNDAMENTAL PARTICLES, ELEMENTS, NANO UNITS, MICRO UNITS, MACRO UNITS, ROCKS, ETC. WHICH ALL OF THESE STRUCTURES HAVE INDEPENDENT MASS UNIT'S SPECIFIC PARTICLE INTELLIGENCE SYSTEM CENTERS (PIS), GRAVITON FORCES, SPECIFIC ENERGY CONTENTS, ALSO EMIT FROM MASSES PARTICLE CLOUDS ORDERS TO CONTROL SURROUNDING OTHER MASS UNITS BY PROPER

USES OF GRAVITON FORCES UNDER PIS DECISIONS AND ORDERS.

THE ELECTRON'S PIS, NEUTRON'S PIS, PROTON'S PIS, NANO UNIT'S PIS, MICRO UNIT'S PIS, MACRO UNIT'S PIS, AND PLANETARY GALACTICAL PARTICLE INTELLIGENCE SYSTEM CENTERS DECIDE, PRODUCE AND EMIT PARTICLE CLOUDS. WHICH THE OTHER ELEMENTS, FUNDAMENTAL PARTICLES, MICRO UNITS, MACRO UNITS AND PLANETARY MASSES UNDER GRAVITON FORCES AND EMITTED CLOUD ORDERS MUST FUNCTION ACCORDING PIS - EMITTED PARTICLE CLOUDS OF DIFFERENT KIND MASS UNITS OF THE UNIVERSE.

UNDER PARTICLE INTELLIGENCE SYSTEMS (PIS) CLOUD ORDERS, THE GRAVITON UNDER ATTRACTIVE AND REPULSIVE FORCES EXECUTE PARTICLE CLOUD ORDERS THROUGH PHYSICAL CHEMICAL MECHANICAL FORCES AS WELL AS UNDER A. S. I. Mol. CIC. AUTONOMOUSLY AND SEQUENTIALLY CONSTRUCT UNIVERSAL OR PLANETARY EXISTING THINGS DIFFERENT CONSTRUCTIONS.

UNDER PIS CLOUD ORDERS, THE GRAVITON UNDER A. S. I. FP. Mol. CIC., A. S. I. N.U. CIC, AND A. S. I. MICRO UNIT CIC. THROUGH ATTRACTIVE GRAVITON FORCE CONTROL CHEMICAL INTERACTIONS, COMBINE AND CONSTRUCT EXISTING THINGS. OR UNDER GRAVITON FORCES BREAK DOWN EXISTING THINGS CONSTRUCTIONS UNDER REPULSIVE FORCES OF GRAVITON, ACCORDING PIS CLOUD ORDERS. THE GRAVITON CONSTRUCTED UNIVERSAL EXISITING THINGS UNDER MOLECULAR EVOLUTION ORDERS AND PATHS.

PARTICLE- MOLECULE NEO GENESIS BY GRAVITON FORCE
First Phase Evolution between Fundamental Particles

FUNDAMENTAL PARTICLE MOLECULE NEO GENESIS UNDER GRAVITON FORCE

THE PRODUCED POST UNIVERSAL EXPLOSION DUST'S FUNDAMENTAL PARTICLES UNDER ATTRACTIVE GRAVITON FORCES ATTRACTED EACH OTHER, COMBINED AND PRODUCED DIFFERENT LARGE FUNDAMENTAL PARTICLE MOLECULE COMPOUNDS CONSTRUCTIONS, UNDER REGENERATIVE A. S. I. FP. Mol. CIC. THIS PHENOMENON IS NEO FUNDAMENTAL PARTICLE COMPOUND MOLECULE GENESIS, WHICH TAKE PLACE UNDER ATTRACTIVE GRAVITON FORCE, THROUGH MOLECULAR EVOLUTION ORDERS. AND CONSTRUCT ENORMOUS DIFFERENT KIND NEW PARTICLE COMPOUNDS.

THROUGH FURTHER USE OF NEWLY PRODUCED FUNDAMENTAL PARTICLE COMPOUNDS, IN CONTINUATION OF EVOLUTION UNDER ATTRACTIVE GRAVITON FORCES COMBINATIONS, THESE NEW MOLECULE COMPOUNDS FURTHER USED TO CONSTRUCT DIFFERENT ELECTRON'S SUBSYSTEM UNITS, NUCLEON'S SUBSYSTEM UNITS, AND

SYSTEM (CONSTRUCTION OF QUARKS) CONSTRUCTIONS. WHICH IN LATER EVOLUTION STAGES CAUSED NEO GENESIS OF ELECTRONS, NEUTRONS, PROTONS, AND ATOMS.

THIS IS MOLECULE EVOLUTION WHICH CREATED DIFFERENT NANO UNITS, MATTER, IN LATER STAGE PRODUCED DIFFERENT MICRO UNITS IN PLANET EARTH UNDER GRAVITON WORK, AND MOLECULE EVOLUTION.

GRAVITON ALSO THROUGH IT'S REPULSIVE FORCES PREVENTED UNWANTED MOLECULE NEO GENESIS FROM PRODUCTION, UNDER DEGENERATIVE A. S. I. FP. Mol. CIC. THE LARGE MOLECULAR COMPOUNDS UNDER REPULSIVE GRAVITON FORCE ORDERS BROKE DOWN INTO SMALL FUNDAMENTAL PARTICLE COMPOUND CONSTRUCTIONS. THROUGH MOLECULAR EVOLUTION ORDERS AND PATHS.

CONSTRUCTION OF NANO UNITS

Element Genesis, Electron - Nucleon Genesis

Atom's Nucleus Genesis, Peripheron Genesis

under Graviton force, Through Evolution

NOW SECOND PHASE, THEREAFTER THIRD, FOURTH, N.th PHASE MOLECULAR EVOLUTION UNDER GRAVITON FORCE

SECOND PHASE MOLECULAR EVOLUTION

PHASE OF PRELIMINARY NANO UNIT GENESIS

AT THIS PHASE THE DIFFERENTLY PRODUCED PARTICLE COMPOUND MOLECULES, PARTICLE CLOUDS, AND FUNDAMENTAL PARTICLES, UNDER ATTRACTIVE GRAVITON FORCES OF MATTER, THROUGH A. S. I. FP. Mol. CIC. COMBINE WITH EACH OTHER, AND PRODUCE DIFFERENT NEW NANO UNIT CONSTRUCTIONS, SUCH AS: NANO UNIT'S SUBSYSTEM UNITS, ELECTRON'S SUBSYSTEM UNITS, PROTON'S SUBSYSTEM UNITS, NEUTRON'S SUBSYSTEM UNITS, ETC.

AT THE SAME TIME, THE UNWANTED FUNDAMENTAL PARTICLE COMPOUND STRUCTUTES UNDER REPULSIVE GRAVITON FORCES, AND DEGENERATIVE A. S. I. FP. Mol. CIC. BREAK DOWN, AND REMOVED FROM INTERNAL ELECTRON, NUCLEON PARTICLE CONSTRUCTIONS. THE SUBSYSTEM UNITS ARE PRELIMINARY INTERNAL ELECTRON, NEUTRON, PROTON, ATOM'S PARTICLE COMPOUND CONSTRUCTIONS.

THE CHEMICAL COMBINATIONS OF DIFFERENT ELECTRONS SUBSYSTEM UNITS, DIFFERENT NUCLEONS SUBSYSTEM UNITS WITH EACH OTHER, UNDER A. S. I. N.U. CIC. PRODUCED LARGER VARIETIES OF DIFFERENT NANO UNIT SYSTEMS CONSTRUCTIONS. SUCH AS PROTON SYSTEMS (QUARKS), NEUTRON SYSTEMS (QUARKS), ELECTRON SYSTEMS, ETC.

AT THIS PHASE DIFFERENT ELECTRON SYSTEMS AND SUBSYSTEM UNITS, NEUTRON AND PROTON SUBSYSTEM UNITS AND SYSTEMS, DIFFERENT MOLECULAR SIZE PARTICLE COMPOUNDS AND FUNDAMENTAL PARTICLE COMPOSITES, DIFFERENT CONSTRUCTION PARTICLE

CLOUDS COMBINED WITH EACH OTHER PRODUCED DIFFERENT LARGER NEW CONSTRUCTIONS NEUTRONS, ELECTRONS, PROTONS, AND OTHER NANO UNIT STRUCTURES.

AT THIS PHASE OF EVOLUTION DIFFERENT NUMBER, DIFFERENT CONSTRUCTION ELECTRONS, NEUTRONS, PROTONS, POSITRONS, ELECTRON NEUTRINOS, ELECTRON ANTINEUTRINOS, ETC. UNDER DIFFERENT REGENERATIVE A. S. I. N. U. Mol. CIC. THROUGH ATTRACTIVE GRAVITON FORCES COMBINED WITH EACH OTHER AND PRODUCED ENORMOUS ANOTHER KINDS DIFFERENT CONSTRUCTIONS, PRODUCED OTHER NANO UNIT CLASSES, AND SUBCLASSES OF NEW CONSTRUCTION ELECTRONS, NEUTRONS, PROTONS, NUCLEI, PERIPHERON, SUBSYSTEM UNITS, ELECTRON NUCLEON SYSTEMS (QUARKS), POSITRONS, ELECTRON NEUTRINO, AND ELECTRON ANTINEUTRINOS, ETC., AT THE SAME TIME NOT NEEDED PARTICLE COMPOUND CONSTRUCTIONS UNDER DEGENRATIVE A. S. I. FP. Mol. CIC. THROUGH REPULSIVE GRAVITON FORCES DESTRUCTED AND REMOVED. THESE ARE PHENOMENONS OF ELECTRON NEO GENESIS, NEO GENESIS OF NEUTRONS, PROTON AND ATOM GENESIS, ETC.

GENESIS OF DIFFERENT ELEMENTS AND ELEMENTAL MOLECULE CONSTRUCTIONS

AT THIS PHASE OF MOLECULAR EVOLUTION DIFFERENT KIND PRODUCED NANO UNITS, SUCH AS DIFFERENT ELECTRONS, NEUTRONS, PROTONS, POSITRONS, ELECTRON NEUTRINO, AND ELECTRON ANTINEUTRINOS, UNDER DIFFERENT COMBINATION, DIFFERENT A. S. I. N.U. CIC.

COMBINED WITH EACH OTHER, AND PRODUCED DIFFERENT KINDS OF THE EXISTING PLANET SPECIFIC ELEMENTS, NANO UNITS. SOME EXAMPLES FROM PRODUCED EARTHS ELEMENTS ARE IN PERIODIC TABLE, AT EARTH MOSTLY ARE BIOFRIENDLY.

GENESIS OF THE BIOMOLECULES UNDER MOLECULAR EVOLUTION

AT THE NEXT PHASE THE CREATED NEW ELEMENTS, AND ELEMENTAL MOLECULAR COMPOUNDS, UNDER REGENERATIVE AUTONOMOUS SEQUENTIAL INTER ATOM BASE MOLECULAR CHEMICAL INTERACTIONS (A. S. I. AB. Mol. CIC.), THROUGH ATTRACTIVE GRAVITON FORCES COMBINED WITH EACH OTHERS, PRODUCED MORE COMPLEX AND LARGE ELEMENTAL MOLECULE COMPOUNDS, LATER BIOMOLECULES CREATED AT PLANET EARTH UNDER EARTH'S MOLECULAR EVOLUTION PHASES.

EVOLUTION IN BIOMOLECULES, GENES
CELLS, TISSUES, ORGANS, AND BODY SYSTEMS
GENESIS OF LIVING THINGS UNDER GRAVITON FORCES

IN THIS PHASE EVOLUTION, UNDER REGENERATIVE AUTONOMOUS SEQUENTIAL INTER BIOMOLECULAR CHEMICAL INTERACTION CHAINS AND CYCLES (REG. A. S. I. B. M. CIC.) UNDER GRAVITON ATTRACTION FORCES, THE PRODUCED DIFFERENT BIOMOLECULES, RNA, DNA, GENES, CHROMOSOMES FROM PREVIOUS PHASES EVOLUTION ERAS COMBINED WITH EACH OTHER, AND PRODUCED THE PRESENTLY EXISTED BIOLOGICAL NANO UNITS, MICRO UNIT, CELLS, TISSUES. ADDITIONALLY, UNDER REPULSIVE

GRAVITON FORCES AND DEGENERATIVE A. S. I. B.M. CIC. THE UNWANTED BIOMOLECULES, CELLS. CHROMOSOMES, GENE'S COMBINATIONS WITH EACH OTHER REPULSED AND REFUSED.

BIOMOLECULAR PHASE EVOLUTION PRODUCED CONSTRUCTION OF ENORMOUS DIFFERENT CHEMICAL FORMULARY CHROMOSOMES, INTERNAL CYTOPLASMIC ORGANS, INTERNAL CYTOPLASMIC SYSTEMS CONSTRUCTIONS. PRODUCED INTERNAL CELL NUCLEI ORGAN CONSTRUCTIONS, AND INTERNAL NUCLEI CELL SYSTEMS CONSTRUCTIONS. DIFFERENT CELL SPECIES PRODUCED.

MICRO EVOLUTION CAUSED CONSTRUCTION OF CELL'S CYTOPLASMIC ORGANS, CELL'S CYTOPLASMIC SYSTEMS, CONSTRUCTION OF MITOCHODRIA, CELL MEMBRANE, CELL'S CYTOPLASMIC PCS, AND PIS CONSTRUCTIONS OF CYTOPLASMA CREATED. MICRO PHASE EVOLUTION CAUSED CONSTRUCTION OF INTERNAL NUCLEI ORGAN CONSTRUCTIONS, AND INTERNAL NUCLEI SYSTEM PRODUCTIONS, SUCH AS CELL DUPLICATION SYSTEMS CONSTRUCTIONS, CELL NUCLEI INTELLIGENCE SYSTEM CENTERS PRODUCTIONS, ETC.

THE HEAVY HIGHER ENERGY CONTENT DOMINANT CELL NUCLEI MICRO UNIT (WHICH HAVE BEEN CONSTRUCTED UNDER PREVIOUS EVOLUTION PHASES), AND POSSESSES STRONG ATTRACTIVE GRAVITON FORCES TAKE NUCLEAR POSITION, AND FROM CELL CENTRAL NUCLEI POSITION ATTRACT RECESSIVE LESSER GRAVITON FORCE CYTOPLASMIC MICRO UNITS, UNDER ATTRACTIVE

GRAVITON FORCES THE NUCLEI AND CYTOPLASMIC MICRO UNIT COMPARTMENTS COMBINE WITH EACH OTHER, UNDER REGENERATIVE AUTONOMOUS SEQUENTIAL INTER MICRO UNIT CHEMICAL INTERACTION CYCLES AND CHAINS, AND PRODUCE BICOMPARTMENT CELL MICRO UNIT CONSTRUCTIONS. THIS IS CELL GENESIS PHENOMENON.

IN WHICH TWO DIFFERENT MICRO UNIT CELL MONO COMPARTMENTS COMBINE WITH EACH OTHERS UNDER GRAVITON ATTRACTIVE FORCES AND PRODUCE ONE BICOMPARTMENT CELL MICRO UNIT STRUCTURE.

THE REVERSE OF THIS CELL GENESIS PHENOMENON, UNDER DEGENERATIVE A. S. I. - MICRO UNIT- CIC. ALSO IS TRUE. UNDER THIS PHENOMENON A GIVEN BICOMPARTMENT CELL CONSTRUCTION, UNDER DEGENERATIVE A. S. I. MICRO- UNIT - CIS, A GIVEN UPLICATE BIOMOLECULAR CONSTRUCTION MICRO UNIT CELL, UNDER REPULSIVE GRAVITON FORCES, THE DUPLICATE MICRO UNIT STRUCTURES AND BIOMOLECULAR CONSTRUCTIONS SEPARATE FROM EACH OTHER, AND TWO DUPLICATE PRODUCED MICRO UNIT CONSTRUCTION REPULSE EACH OTHER AND SEPARATE.

ABOVE REPULSIVE PHENOMENON HAVE DIFFERENT ANOTHER PROBABILITIES, WHCH HAVE BEEN EXPLAINEED IN OTHER BOOKS OF AUTHOR.

WHEN A DOMINANT HIGH WEIGHT, HIGH GRAVITON FORCE CELLS NUCLEUS COMBINE WITH RECESSIVE LESS WEIGHT, LESS ENERGY GRAVITON FORCE CYTOPLASM, ALWAYS CELL NUCLEUS TAKE DOMINANT CENTRAL POSITION UNDER DOMINANT HIGH FORCE GRAVITON, AND LESS WEIGHT

CYTOPLASM WITH RECESSIVE LESSER QUANTITY GRAVITON FORCE STAY AT PERIPHERY. UNDER AUTONOMOUS SEQUENTIAL INTER MICRO UNIT CHEMICAL INTERACTION CHAINS AND CYCLES BOTH COMBINE WITH EACH OTHER.

FUNDAMENTAL PARTICLE'S BIOLOGICAL SCIENCE IN PLANTS
FUNDAMENTAL PARTICLES BIOLOGICAL CHEMISTRY IN PLANTS

Plant's Electron, Nucleon Genesis, Atom Genesis, Cell Genesis, Color Neo Genesis in Plants

THE NEO- PARTICLE COMPOUND GENESIS INSIDE PLANT'S ELECTRONS AND NUCLEONS IS COMBINATION OF EXOGENOUS AIRBORNE LIGHT FUNDAMENTAL PARTICLES, WITH PLANT'S INDIGENOUS INTER ELECTRON NUCLEON FUNDAMENTAL PARTICLE COMPOUNDS. AND NEO GENESIS OF NEWLY CONSTRUCTED LIGHT FUNDAMENTAL PARTICLE COMPOUNDS CONSTRUCTION, AND PRODUCTION OF NEW ELECTRONS, NEUTRONS, PROTONS, ATOMS, AND NEW CELL'S CONSTRUCTIONS. THESE PHENOMENONS ARE NEO GENESIS OF ELECTRONS AND NUCLEONS, AS WELL AS NEO GENESIS OF ATOMS AND CELLS IN PLANTS.

THE AIRBORNE EXOGENOUS SUN ORIGIN LIGHT FUNDAMENTAL PARTICLES (EX. - Y- FP), SHINE DIRECT ON PLANT'S ELECTRONS AND NUCLEONS, THE EX. —Y –FP ENTER

INTO PLANT'S INTERNAL ELECTRON, NUCLEON SUBSYSTEM-UNIT'S CHEMICAL LAB.S. THE EXOGENOUS LIGHT FUNDAMENTAL PARTICLES COMBINE WITH INIGENOUS INTER ELECTRONS NUCLEON'S FUNDAMENTAL PARTICLE COMPOUNDS (FP – COMP.) AND PRODUCE LIGHT FUNDAMENTAL PARTICLE COMPOUNDS (Y - FP - COMP.), AND PRODUCED LIGHT FUNDAMENTAL PARTICLE COMPOUNDS CONSTRUCT PLANT'S INTERNAL ELECTRON, INTERNAL NUCLEON PARTICLE COMPOUND CONSTRUCTIONS THROUGH A. S. I. F.P. Mol. C.I.C.

 THIS IS PHENOMENON OF NEO- PARTICLE COMPOUND GENESIS, IT IS PROCESS OF NEO-ELECTRON CONSTRUCTION WITH LIGHT FUNDAMENTAL PARTICLE COMPOUNDS, IT IS NEO NUCLEON GENESIS AND NEO ATOM GENESIS WITH PLANET SUN ORIGIN DIFFERENT COLOR EXOGENOUS LIGHT FUNDAMENTAL PARTICLES. AS WELL AS IT IS NEO CELL GENESIS OF PLANTS WITH THE COLOR OF THE LIGHT PARTICLES.

DEPENDING WHAT COLOR EXOGENOUS LIGHT PARTICLES COMBINE WITH PLANTS INTERNAL ELECTRON NUCLEON PARTICLE COMPOUNDS, UNDER A. S. I. F.P. Mol. C.I.C., THE PRODUCED INTERNAL ELECTRON-NUCLEON LIGHT PARTICLE – COMPOUNDS COLORS WILL BE EXACTLY THE SAME COLOR AS THE COMBINING EXOGENOUS LIGHT PARTICLES COLORS, IF COMBINING EXOGENOUS LIGHT PARTICLE IS GREEN, THE PARTICLE COMPOUNDS COLORS WILL TINT GREEN AS WELL, THE RED COLOR OR YELLOW COLOR PETALS COMBINING LIGHT FUNDAMENTAL PARTICLES COLORS, RESPECTEDLY

ARE RED – LIGHT PARTICLES AND YELLOW LIGHT PARTICLES COLORS.

THIS PHENOMENON IS THE CAUSE FOR PLANTS ELECTRONS, NEUTRONS, CELLS, ATOMS, COLOR CHANGES AFTER THE COMBINATIONS WITH EXOGENOUS FUNDAMENTAL PARTICLES, THE PLANT CELLS TINT WITH COMBINING LIGHT PARTICLES COLOR, THE PARTICLE COMPOUND COLORS CHANGES ALSO INTO COMBINING LIGHT PARTICLES COLORS.

THIS IS PHENOMENON OF COLOR – GENESIS IN PLANTS, IF COMBINING LIGHT PARTICLE IS GREEN COLOR THEREFORE PLANT ELECTRON-NUCLEONS PARTICLE- COMPOUNDS CELLS GETS GREEN COLOR AS WELL, (OR COLOR- GENESIS PHENOMENON IN PLANTS).

EARTH'S SENSIBLE LIGHT FUNDAMENTAL PARTICLES

MAJORITIES OF SUN- ORIGIN BIOFRIENDLY LIGHT FUNDAMENTAL PARTICLES CLASSES, ARE NON SENSIBLE AND NOT - DISCOVERED LIGHT FUNDAMENTAL PARTICLES (BF- NS- ND –Y. FP). BIOFRIENDLY NON- SENSIBLE LIGHT PARTICLE CLASSES CONSTRUCT MOST OF PLANET EARTH'S INTER ELECTRONS - NUCLEONS PARTICLE COMPOUND CONSTRUCTIONS, ALSO INTERNAL MONO- ATOMS STRUCTURES, AND NANO-UNIT CONSTRUCTIONS, NANO- UNIT CLOUDS, AND ATOM – CLOUDS.

EARTH'S DISCOVERED SENSIBLE LIGHT FUNDAMENTAL PARTICLES CLASSES, PRESENTLY ARE KNOWN AS LIGHT- WAVES, THE SENSIBLE LIGHT FUNDAMENTAL PARTICLES AT

EARTH ARE: GREEN –COLOR LIGHT PARTICLES, RED – LIGHT PARTICLES, AND YELLOW COLOR LIGHT PARTICLE CLASSES, FROM OTHER SENSIBLEE PARTICLES ARE, BLUE COLOR FUNDAMENTAL PARTICLES, OR VIOLET COLOR LIGHT PARTICLE CLASSES, WHICH ALL OF THESE PARTICLES ALSO CONSTRUCT PLANET EARTHS ATOM'S LECTRON – NUCLEON PARTICLE COMPOUND CONSTRUCTIONS.

THERE ARE ADITIONALLY LARGE NUMBERS OF SENSIBLE DIFFERENT OTHER LIGHT FUNDAMENTAL PARTICLES OF DIFFERENT COLORS WHICH ARE NOT SEEN BEFORE BY MAN KINDS. THESE DIFFERENT CLASSES OF SENSIBLE PARTICLES WHICH CONSTRUCT CRYSTALLINE OR OTHER KIND ATOMS OF EARTH PLANET, AND EXST IN LIGHT PARTICLE COMPOUND CONSTRUCTION FORM BUILDING STRUCTURES OF DIFFERENT ELECTRONS NUCLEONS MOLECULAR CONSTRUCTIONS. THESE PARTICLES CAN BE RELEASED THROUGH DEGENERATIVE AND REGENERATIVE A. S. I. FP. Mol. CIC. AS FREE SENSIBLE LIGHT FUNDAMENTAL PARTICLES WITH DIFFERENT BRILLIANT INTERESTING COLORS WHICH VARIES FROM ONE CLASS TO THE OTHER, AND FURTHER STUDIED.

FUNDAMENTAL PARTICLE BIOCHEMISTRY

THE COLOR OF PLANT'S FUNDAMENTAL PARTICLE COMPOUNDS ARE WATER CLEAR WHITE COLOR, WHEN PLANT'S ELECTRON, NEUTRON, PROTNS, ATOMS AND CELLS NOT COMBINED WITH LIGHT-PARTICLES.

IN CASES WHEN PLANTS TISSUE'S, POSSESS NO CHEMICAL COMBINATIONS WITH AIRBORNE LIGHT- FUNDAMENTAL PARTICLES, MOSTLY IN THESE CASES, THE CELL'S COLORS, THE ATOM'S COLOR, THE PLANT'S TISSUES COLOR ALL STAY COLORLESS, OR WATER- CLEAR WHITE COLOR, IN MOST TISSUES WHICH, THESE CELLS DO NOT POSSESS, OR MINIMALLY POSSESS LIGHT PARTICLES COMBINATIONS WITH INTER ELECTRON – NUCLEON PARTICLE COMPOUNDS, AND THERE IS ONLY MINIMAL OR, LESS LIGHT PARTICLE COMPOUNDS INSIDE THEIR PLANT'S ELECTRONS NUCLEONS.

THESE CASES MOSTLY OCCUR AT EARLY STAGES, WHEN IN THIGHTLY CLOSED THICK FLOWER BUDS, OR INSIDE THICK NIDUS, WHEN THE LIGHT FUNDAMENTAL PARTICLES ARE UNABLE TRANSIT INTO INSIDE UN-PENETRABLE CLOSED PLANT SPACES.

AT EARLY GROWTH STAGE OF THIGHTLY CLOSED THICK BUDS, WHICH DOES NOT ALLOW THE LIGHT PARTICLES INTER INTO THIGHTLY CLOSED THICK BUDS, AND DOES NOT ALLOW LIGHT PARTICLES CHEMICAL COMBINATIONS WITH INTER ELECTRONS NUCLEONS PARTICLE – COMPOUNDS TO TAKE PLACE, THIS PHENOMENON ALSO OCCUR INSIDE HARD SHELLS OF NUTS AT EARLY STAGE GROWTH, WHICH LIGHT PARTICLES IS UNABLE CROSS THICK SHELLS AND COMBINE WITH PIT CELL'S ELECTRONS NUCLEONS, THESE TISSUES STAY WHITE COLORLESS CLEAR BECAUSE TISSUES ELECTRONS NUCLEONS POSSESS NO COMBINATIONS WITH LIGHT PARTICLES.

PARTICLE BIOCHEMISTRY

PLANT'S ATOM GENESIS, PLANT'S CELL GENESIS, WITH GREEN LIGHT PATICLE COMBINATIONS

PLANT'S ELECTRON – PROTON – NEUTRON FUNDAMENTAL PARTICLE COMPOUND GENESIS,

WITH GREEN COLOR LIGHT FUNDAMENTAL PARTICLE COMBINATIONS:

IN EARLY SPRING, WHEN THE EARLY PLANT STEM CELLS MULTIPLY WITH EXPONENTIAL SPEEDS, AND IN MANY SPECIES THE EARLY FLOWER'S THICK BUDS ARE THIGHTLY CLOSED, THE LIGHT PARTICLES ARE NOT CAPABLE FOR DIRECT ENTERANCES INTO INSIDE THE THIGHTLY CLOSED CUSPIDS, IN THESE CASES AT EARLY STAGE BUDS, THE ALL PETALS, OVARIES, STAMEN, UNDER THESE CONDITIONS (LOCATED INSIDE THIGHTLY CLOSED CUSPIDS, NON PENETRABLE TO LIGHT PARTICLES) ALL FLOWER STAY COLORLESS WHITE, EXCEPT EXTERNAL CUSPIDS WHICH HAS DIRECT EXPOSURE TO GREEN AIRBORNE LIGHT FUNDAMENTAL PARTICLES, WHICH TINT GREEN, BECAUSE OF GREEN LIGHT PARTICLE COMBINATION.

IN THIGHTLY CLOSED BUDS WITH THICK CUSPIDS ONLY THE DOMINANT GY – FP IS CAPABLE TRANSIT THICK CUSPIDS INTO INSIDE CLOSED CUSPIDS. AND COMBINE WITH INTERNAL ELECTRONS NUCLEONS PARTICLE COMPOUNDS OF THE PETALS, OVARIES AND STAMEN INSIDE SUBSYSTEM UNITS CHEMICAL LAB.S UNDER A. S. I. FP. Mol. CIC. AND

PRODUCE GREEN LIGHT FUNDAMENTAL PARTICLE COMPOUNDS (GY- FP – COMPOUNDS) CONSTRUCTIONS FLOWER PETALS, OVARIES AND STAMENS WHEN THE CUSPIDS ARE THIGHTLY CLOSED, AND CAUSE THE COLOR OF FLOWERS INSIDE CLOSED CUSPIDS BECOMES ALL GREEN COLOR PETALS, GREEN COLOR STAMENS AND GREEN COLOR OVARIES.

IN FIRST PHASE THE EXOGENOUS AIRBORNE LIGHT FUNDAMENTAL PARTICLES, EXCEPT THE DOMINENT GREEN LIGHT FUNDAMENTAL PARTICLES ARE NOT CAPABLE TO TRANS – CROSS THE CLOSED CUSPIDS THICK WALLS, AND THIS CUSPID CLOSURE IS THE BEST PROTECTION PROCEDURE FOR OVARIES, STAMEN, PETALS OF THE EARLY STAGE FLOWER'S STEM CELLS, WHICH PROTECTS THESE INTERNAL FLOWER STRUCTURES FROM EXTERNAL FROST BITE, AND COLD WEATHER CHANGES OR OTHER HAZARDS.

AT THIS STAGE ONLY DOMINANT GREEN LIGHT PARTICLES CURRENTS, ARE CAPABLE DIRECTLY TO TRANS-CROSS THICK CUSPIDS, THROUGH P.A.S. –PARTICLE CIRCULATION SYSTEMS INTO INSIDE BUD'S CLOSED CAVITY, AND COMBINE WITH INTERNAL ELECTRON-NUCLEON PARTICLE COMPOUNDS, AND PRODUCE GREEN LIGHT PARTICLES – COMPOUNDS CONSTRUCTIONS INSIDE ELECTRON'S- NUCLEON'S OF PETALS, OVARIES, STAMENS, WHEN THE BUDS, CUSPIDS ARE CLOSED. AT THIS STAGE, THE PETALS, STAMENS, OVARIES PARTICLE COMPOUND CONSTRUCTIONS MOSTLY ARE CONSTRUCTED WITH GREEN LIGHT PARTICLE COMPOUNDS, ENTIRE FLOWERS CUSPIDS, PETALS, OVARIES, AND STAMEN ALL ARE GREEN COLOR.

THIS IS PHENOMENON OF PARTICLE COMPOUND NEO- GENESIS, THE ELECTRON NEO- GENESIS, THE NUCLEON NEO -GENESIS PHENOMENON.

AT THIS STAGE THE OTHER LIGHT FUNDAMENTAL PARTICLES ARE RECESSIVE, THEY ARE NOT CAPABLE TO TRANSIT THROUGH P.A.S. CURRENTS CIRCULATION SYSTEMS, INTO INSIDE CAVITY OF THIGHTLY CLOSED, THICK CLOSED CUSPIDS, (ONLY DOMINANT GY –FP POSSESS THIS CAPABILITY).

PLANT GENESIS AND ANIMAL GENESIS

ALTERNATING PLANT'S STEM CELL'S MUTATIONS WITH MEDIA CHEMICAL FORMULARY CHANGES IS THE CAUSE OF PLANT GENESIS AND STEM CELL DIFFERENTIATIONS

PARTICLE BIOCHEMISTRY, NANO UNIT AND MICRO UNIT BIOCHEMISTRY

THE PLANT'S ALTERNATING STABLE STEM CELL'S MUTATION WITH ALTERNATING MEDIA CHEMICAL FORMULARY CHANGES IN EMBRYONIC STAGE - PLANTS, EQUALS TO PLANT GENESIS AT PRESENT TIMES AS WELL IT IS EQUAL TO PLANT GENESIS DURING THE

MOLECULAR EVOLUTION OF THE PAST WORLD HISTORY.

CONSTRUCTION OF ANIMAL FETUSES IN UTERUS

(FETUS GENESIS IN ANIMALS)

INTRAUTERINE ALTERNATING STABLE STEM CELL MUTATION SEQUENCES (A. S. S. C. M. S.), WHEN ONE STABLE MUTATION SEQUENCE ALTERNATE WITH ONE STABLE MEDIAL CHEMICAL FORMULARY CHANGE SEQUENCES (S. M. C. F. C. S.), AUTONOMOUSLY, SEQUENTIALLY, WITH EXPONENTIAL SPEEDS OF MULTIPLICATIONS OF EMBRYONIC STEM CELLS AND REPRODUCTIONS OF NEW TISSUES WITH NEW PRODUCT STEM CELLS AND NEW MEDIAS SEQUENTIALLY ONE AFTER THE OTHER, IS CAUSES OF NEO- GENISIS OF NEW TISSUES, NEW ORGANS AND NEW SYSTEMS, FINALLY IS CAUSES OF PRODUCTS OF NEWBORN ANIMAL GENESIS, AND NEW BORN PLANT GENESIS. WHICH TODAY'S LIVING THING GENESIS IS EQUAL TO MOLECULE EVOLUTION NEO LIVING THING GENESIS PROCESSES.

WHEN ONE STABLE MUTATION SEQUENCE FOLLOW THE OTHER STABLE MEDIA CHEMICAL FORMULARY CHANGE SEQUENCE, WITH EXPONENTIAL MUTATION SPEEDS OF STEM CELLS ONE AFTER THE OTHER.

(FETAL STEM CELL MUTATION IN EMBRYONIC PHASES ALTERNATE WITH FIX MEDIA

CHEMICAL FORMULARY CHANGES, THIS PROCESS IS EQUAL TO EVOLUTION PATH).

WHEN RAPIDLY ONE STABLE MUTATION STEM CELL, MUTATE INTO ANOTHER STABLE MUTATED STEM CELL WITH EXPONENTIAL SPEED OF INTERACTIONS UNDER A. S. I. N.-U. –C.I.C., THIS PHENOMENON CAUSE RAPID HIGH SPEED MEDIA CHEMICAL FORMULARY CHANGES WITH SAME EXPONENTIAL SPEEDS OF SEQUENTIAL MUTATIONS, ONE BY ONE.

THAT MEANS ONE STABLE MUTATIONS CAUSE ON STABLE MEDIA CHEMICAL FORMULARY CHANGES, AND ONE MEDIA CHEMICAL FORMULARY CHANGES PRODUCE ANOTHER STABLE STEM CELL MUTATIONS, AND SO ON. THIS PHENOMENON IS THE CAUSE FOR CELL DIFFERENTIATIONS IN PLANTS AND ANIMALS EMBRYONIC STATES SEQUENTIAL SPEED STABLE STEM CELL'S REPRODUCTIONS, AND FETUS BODY GENESIS, THROUGH RAPID EXPONENTIAL SPEED STEM CELL MUTATION AND ORAGN CONSTRUCTION PHENOMENONS.

IN UTERO BODY GENESIS PHENOMENONS ARE EQUALS TO EVOLUTION PATH MOLECULAR GENESIS ORDERS AND PATHS. ADDITIONALLY, IN UTERO FETAL GENESIS IS ALSO EQUAL TO TURN OVER PROCESS BODY GENESIS PHENOMENON AS WELL. AND ALL THREE DIFFERENT PHENOMENONS ARE EQUAL SEQUENCES.

THESE THREE DIFFERENT PROCESSES ALL CAN BE USED AS GUIDE LINES FOR TISSUE GENESIS AT FUTURES CHEMICAL LAB.S FOR NEW BODY ORGAN CONSTRUCTION, REPRODUCTION AND IMPLANTS IN FUTURE LAB.S. (THE

FETAL STAGES SEQUENTIAL ALTERNATING MUTATION – MEDIA CHANGES SEQUENCES, ARE EQUAL TO EVOLUTION PHASES STABLE SEQUENTIAL MEDIA – MUTATION SEQUENTIAL CHANGE ALTERNATIONS).

NEO TISSUE GENESIS IN PLANTS

SEQUENTIAL STEM CELL MUTATION PHASES, ALTERNATING WITH MEDIA CHANGE PHASES

WHEN THE ONE PHASE STABLE STEM CELL MUTATE AND ALTERNATE ONE SEQUENCE STEM CELLS MEDIA CHEMICAL FORMULARY CHANGES, THIS IS ONLY ONE CYCLE AUTONOMOUS SEQUENTIAL STABLE MUTATION. BUT IN ANIMAL AND PLANT EMBRYONIC STAGE THESE SEQUENCES CONTINUE IN MULTIPLE PHASES SEQUENTIAL CYCLES, ONE CYCLE FOLLOW THE PREVIOUS ONE, ONE STEM CELL MUTATIONS FOLLOW THE NEXT ONE STBLE MUTATIONS IN MULTIPLE ROW OF STABLE MUTATION SEUENCES AND PRODUCE DIFFERENT MULTI SEQUENTIAL AUTONOMOUS NEW SPECIES STEM CELLS GENESIS AND NEW STEM CELL DIFFERENTIATIONS ONE AFTER THE OTHER SEQUENTIALLY AND AUTONOMOUSLY WITH PARTICLE ACCURACY PRECISIONS CREATE NEW BODY ORGANS AND NEW BODY SYSTEMS.

THIS PROCESS IS CAUSE OF NEW TISSUE-GENESIS (STABLE CELL MUTATIONS AND CELL DIFFERENTIATION), THIS IS CAUSE OF NEO BODY ORGAN GENESIS (NOG), NEO BODY SYSTEM GENESIS (NSG), AND NEW FETUS CONSTRUCTION

IN ANIMALS (NAG) AND PLANT EMBRYONIC STATES OR NEW PLANT GENESIS (NPG).

THIS IS PHENOMENON OF COMPLETE FETUS –GENESIS IN EMBRYO, UNDER EXPONENTIALLY RUNNING AUTONOMOUS SEQUENTIAL INTER MOLECULAR AND BIOMOLECULAR CHEMICAL INTERACTION CYCLES AND CHAINS, WHICH CAUSING AUTONOMOUS SEQUENTIAL ALTERNATING STEM CELL MUTATION SEQUENCES, AND IN EACH SEQUENCE CAUSING MEDIA CHEMICAL FORMULARY CHANGES ALTERNATE WITH STABLE MUTATION PHASES (ONE BY ONE FOR EACH PHASE OF MUTATION).

THE ALTERNATING CHEMICAL FORMULARY CHANGES PRODUCE DIFFERENT TISSUES AND CELLS SPECIES (CAUSE CELL DIFFERENTIATIONS), PRODUCE DIFFERENT BODY ORGAN CONSTRUCTIONS AND DIFFERENT BODY SYSTEM CONSTRUCTIONS. THIS IS THE CAUSE HOW ANIMALS AND PLANTS BODY CONSTRUCT A WHOLE CREATURE. (NOG, NSG. NAG, NPG).

NEO PLANT GENESIS WITH ATOM MUTATION

Tissue Genesis and cell differentiation in early stage Stem Cells Mutations

MOLECULAR EVOLUTION IS EQUAL TO PRESENT TISSUE DIFFERENTIATIONS

Combination of Flowers indigenous internal cell's Electron, Nucleon Particle Compounds with airborne Particles produce new Electron Nucleon Genesis, and new Flower Cell Genesis (Stable Mutations).

AT EARLY SPRING DURING PLANT STEM CELL RAPID GROWTH STAGES, THE LEAF'S INTERNAL CELL ELECTRONS, NUCLEONS PARTICLE COMPOUNDS COMBINE WITH EXOGENOUS AIRBORNE SUN ORIGIN GREEN COLOR LIGHT FUNDAMENTAL PARTICLES THROUGH REGENERATIVE A. S. I. FP. Mol. CIC. AND PRODUCE GREEN LIGHT PARTICLE COMPOUND ELECTRONS AND NUCLEONS FOR PLANT'S LEAFS AND STEMS. IN CONTARY THE ROOT CELL'S ELECTRONS AND NUCLEONS DO NOT HAVE DIRECT AIRBORNE GREEN LIGHT PARTICLE EXPOSURES, THE ROOTS CELLS ATOMS PARTICLE COMPOUNDS DO NOT COMBINE WITH GREEN LIGHT PARTICLES, AND DO NOT PRODUCE GREEN COLOR ELECTRONS, NUCLEONS PARTICLE COMPOUNDS ATOMS CONSTRUCTIONS AND GREEN COLOR CELL GENESIS.

IN PLANTS THE MEDIA CHEMICAL FORMULARY CHANGES OF LEAVES AND STEMS ELECTRONS NUCLEONS, CAUSE PLANTS STEM CELLS MUTATE AND CHANGE TO OTHER COLOR STEM CELLS (STABLE MUTATION). THE MEDIA CHEMICAL FORMULARY CHANGES CAUSE STEM CELLS MUTATIONS TO ANOTHER STRUCTURES.

FOR EXAMPLE, IN EARLY PHASE ZYGOTES GROWTH IN PLANT SEEDS, THE EARLY STAGE ROOT CELLS UNDER GRAVITON FORCES GROW DEEP IN SOIL, AND THEIRE

MEDIA IS SOIL, AND THEY DO NOT COMBINE WITH AIRBORNE PARTICLES, THEY MUTATE DIFFERENT THAN THE LEAFS AND STEMS MUTATIONS, WHICH THE LEAF'S MEDIA CHEMICAL FORMULARY IS AIR FUNDAMENTAL PARTICLES, AND GREEN LIGHT FUNDAMENTAL PARTICLES COMBINATIONS UNDER A. S. I. FP. Mol. CIC. WILL PRODUCE DIFFERENT COLOR LIGHT PARTICLE COMPOUND ATOM CONSTRUCTIONS IN THE LEAFS, IN CONTRAST TO THE ROOTS.

MEDIA CHEMICAL FORMULARY CHANGE PRODUCE
DIFFERENT ATOM CONSTRUCTIONS

OPPOSITE OF ABOVE INTERACTIONS ALSO IS TRUE, THE CREATED NEW STEM CELLS MUTATION PRODUCE DIFFERENT KIND STEM CELLS MEDIA, AND THIS SECONDARY MEDIA CHEMICAL FORMULARY CHANGE CAUSE TIRTIARY EMBRYONIC STEM CELLS STABLE MUTATIONS, WHICH THEY GROW SEQUENTIALLY AND AUTONOMOUSLLY UNDER BIOMOLECULAR CHEMICAL INTERACTIONS AND PRODUCE CELL DIFFERENTIATIONS, NEW TISSUE GENESIS, NEW BODY ORGAN GENESIS, NEW SYSTEM CONSTRUCTIONS. THESE AUTONOMOUS SEQUENTIAL INTERBIOMOLECULAR INTERACTIONS PROCEED WITH EXPONENTIAL SPEED OF STEM CELL MULTIPLICATIONS. AND PRODUCTION OF NEWBORN ANIMAL OR PLANT.

ATOM MUTATION CAUSE STEM CELL MUTATION

SEQUENTIAL GENESIS OF DIFFERENT TISSUE SPECIES

COMBINATION OF INTERNAL STEM CELL ATOM'S PARTICLE COMPOUNDS

WITH DIFFERENT EXOGENOUS Y - FP PRODUCE DIFF. STEM CELL MUTATION

FOLLOWING WIDELY OPENING CUSPIDS, THE FLOWERS INTERNAL PETALS, STAMEN AND OVARIES ELECTRON - NUCLEON PARTICLE COMPOUNDS COMBINE WITH INCOMING AIRBORNE DIFFERENT SENSIBLE LIGHT FUNDAMENTAL PARTICLES, AND PRODUCE DIFFERENT CONSTRUCTION LIGHT FUNDAMENTAL PARTICLE COMPOUNDS INSIDE THE INTERNAL ELECTRONS AND NUCLEONS MOLECULAR CONSTRUCTIONS OF THE FLOWER'S PETAL, STAMEN AND OVARIES. THE LIGHT FUNDAMENTAL PARTICLE COMPOUND CONSTRUCTED ELECTRONS NUCLEONS TINT WITH LIGHT FUNDAMENTAL PARTICLES COLORS, THE COLORS OF THE PETALS, STAMEN, OVAIRIES CHANGE ACCORDING THE COMBINING LIGHT FUNDAMENTAL PARTICLE MIXTURE COLORS. MUTATION OF THE PLANTS CELLS ELECTRONS NUCLEONS AND PARTICLE COMPOUNDS CONSTRUCTIONS INTO ANOTHER STRUCTURES AND COLORS. CAUSES MUTATIONS OF THE PETALS, OVARIES, STAMENS PLANTS CELL CHANGES INTO ANOTHER COLOR ACCORDING THE LIGHT PARTICLE COMPOUND CONSTRUCTIONS. THE ATOM MUTATIONS CAUSES THE CELLS AND TISSUE MUTATIONS IN CONSEQUENCE.

THIS IS STEM CELL'S MUTATIONS (NEO STEM CELL GENESIS) SECONDARY TO ATOM MUTATIONS PHENOMENON).

WHEN THE CUSPIDS ARE CLOSED, PETALS, STAMEN, OVARY PARTICLE COMPOUNDS COMBINED WITH DOMINANT GREEN COLOR FUNDAMENTAL PARTICLES HAVE GREEN COLORS, WHEN CUSPIDS OPENED, FLOWERS MUTATED TO OTHER STABLE COLORS FOLLOWING A. S. I. FP. Mol. CIC. AND COMBINATIONS WITH ANOTHER COLOR INCOMING EXOGENOUS AIRBORNE LIGHT FUNDAMENTAL PARTICLES.

FOLLOWING WIDELY OPENING FLOWER CUSPIDS, DIFFERENT AIRBORNE COLOR LIGHT FUNDAMENTAL PARTICLES (SUCH AS Yr. –FP, Y. y. –FP, Y. b. –FP, Y. v. - FP, ETC.) DIRECTLY ENTER INTO SUBSYSTEM UNITS LAB.S OF FLOWER'S ELECTRONS AND NUCLEONS, UNDER REGENERATIVE A. S. I. FP. Mol. CIC. COMBINE WITH PARTICLE COMPOUNDS OF PETALS, OVARIES, STAMENS, INTERNAL ELECTRON NUCLEON PARTICLE COMPOUNDS, PRODUCE DIFFERENT COLOR YELLOW, RED, VIOLET, BLUE COLOR, DIFFERENT FUNDAMENTAL PARTICLE COMPOUNDS, AND CONSTRUCT THE FLOWERS ATOMS PARTICLE COMPOUND CONSTRUCTIONS WITH DIFFERENT COLOR RED, YELLOW, BLUE, VIOLET LIGHT PARTICLES COMPOUND CONSTRUCTIONS, THE COLOR OF THE FLOWERS CELLS TINT TO THE SAME LIGHT FUNDAMENTAL PARTICLE'S COLORS. ABOVE PHENOMENON IS ANOTHER TYPES ALTERNATING STABLE MUTATIONS FROM ONE KIND TO OTHER.

SEQUENTIAL STABLE MUTATIONS (SERIAL MULTI - MUTATIONS) UNDER A. S. I. FP. Mol. CIC. ANOTHER STABLE ALTERNATING CELL MUTATIONS IN PLANT'S FLOWERS

BEFORE OPENING CUSPIDS, WHEN THE THICK CUSPID ARE THIGHTLY CLOSED, DIFFERENT RED, YELLOW, BLUE, VIOLET COLORED LIGHT FUNDAMENTAL PARTICLES CAN NOT TRANSIT INTO CLOSED CUSPIDS, AND AT THIS STAGE THE COLOR LIGHT PARTICLES CAN NOT COMBINE WITH FLOWER'S ELECTRONS AND NUCLEONS COMPOUNDS. AT THIS STAGE ONLY DOMINANT GREEN LIGHT FUNDAMENTAL PARTICLES ARE CAPABLE TO TRANSIT THROUGH PAS – PCS INTO INTERNAL SUBSYSTEM UNITS CHEMICAL LAB.S OF THE PETALS, OVARIES, AND STAMENS INTERNAL ELECTRON AND NUCLEON'S LAB.S.

THE GREEN LIGHT PARTICLES UNDER A. S. I. FP. Mol. CIC. COMBINE AND PRODUCE GREEN COLOR LIGHT FUNDAMENTAL PARTICLE COMPOUNDS, PRODUCE GREEN COLOR PETALS, GREEN COLOR STAMENS, GREEN COLOR OVAIRES. WHEN THE CUSPIDS STILL ARE TIGHTLY CLOSE AND IMPENETRABLE TO THE OTHER PARTICLES. THIS IS PHENOMENON OF PRODUCTION OF NEW KIND STABLE MUTATIONS UNDER THE REGENERATIVE A. S. I. FP. Mol. CIC. WHEN FLOWER CUSPIDS ARE CLOSED.

AT THE SECOND STAGE WHEN THE FLOWER CUSPIDS OPEN WIDELY, THE YELLOW, RED, BLUE, VIOLET DIFFERENT LIGHT

FUNDAMENTAL PARTICLE DIRECT AIRBORNE ENTER INTO FLOWER PETALS, STAMEN, OVARIES INTERNAL ELECTRON NUCLEON SUBSYSTEM UNITS CHEMICAL LAB.S, UNDER A. S. I. FP. Mol. CIC. COMBINE WITH FLOWERS INTER ELECTRON NUCLEON PARTICLE COMPOUNDS, AND PRODUCE DIFFERENT COLOR LIGHT FUNDAMENTAL PARTICLE COMPOUND CONSTRUCTIONS, AND CHANGE IN FLOWER TISSUES PARTICLE MOLECULAR COMPOUND CONSTRUCTIONS INTO RED LIGHT PARTICLE COMPOUND CONSTRUCTIONS, YELLOW, BLUE, VIOLET COLOR FUNDAMENTAL PARTICLE COMPOUND CONSTRUCTIONS, OR MIXED COLOR LIGHT FUNDAMENTAL PARTICLE COMPOUND CONSTRUCTIONS.

THE FLOWERS, OVARIES, STAMENS, PETALS COLOR CHANGE FROM GREEN INTO ANOTHER DIFFERENT NEW COLORS CELLS, ATOMS, ELECTRONS, NUCLEONS. THIS IS PHENOMENONS OF GENESIS OF NEW ALTERNATING STABLE CELL MUTATIONS, FROM ONE KIND TO ANOTHER KIND. FROM GREEN COLOR STABLE MUTATIONS TO ANOTHER DIFERENT COLOR STABLE MUTATIONS.

THIS IS MUTATION OF STEM CELLS FROM ONE LIGHT PARTICLE COMPOUND, TO ANOTHER PARTICLE COMPOUND. THIS PHENPMENON IS NEO- STEM CELL GENESIS UNDER A. S. I. F.P. Mol. C.I.C., THIS PHENOMENON IS CLEARLY SHOWING HOW A. S. I. FP. Mol. CIC. ARE THE CAUSE OF MUTATIONS OR CHANGES OF THE MOLECULAR CHEMICAL COMPOUND CONSTRUCTIONS FROM ONE STABLE STEM CELL MUTATION TO OTHER. PHENOMENON OF SEQUENTIAL MULTI - MUTATIONS.

ALSO THIS PHENOMENON CLEARLY SHOW DIFFERENT STRUCTURAL MOLECULAR CONSTRUCTIONS (SUCH AS DIFFERENT CELL MUTATION OR DIFFERENT ATOM MUTATIONS) PRODUCE DIFFERENT PPHYSICAL CHEMICAL AND BIOLOGICAL FUNCTIONS. (DIFFERENT CELL SPECIES AS WELL AS DIFFERENT ATOM SPECIES HAVE CHEMICAL STRUCTURE CONSTRUCTION SPECIFIC MOLECULAR FUNCTIONS).

"ONE CIRCUIT (ARCH) – PCS BETWEEN: C-PIS, S-PIS, M-PIS"

THE INTELLIGENT PLANTS, AND NON – INTELLIGENT PLANTS

PLANTS CENTRAL PARTICLE INTELLIGENCE SYSTEMS CENTERS (C – PIS)

PLANTS PERIPHERAL PARTICLE INTELLIGENCE SYSTEM CENTERS (S – PIS + M – PIS)

SOME PLANT SPECIES DO NOT HAVE PARTICLE INTELLIGENCE SYSTEMS, THERE ARE OTHER PLANTS SPECIES POSSESSES PARTIAL AND INCOMPLETE CENTRAL PARTICLE INTELLIGENCE SYSTEM CENTERS, AND MANY OTHER PLANT SPECIES HAVE BOTH CENTRAL PARTICLE INTELLIGENCE SYSTEM CENTERS, AND PERIPHERAL PARTICLE INTELLIGENCE SYSTEM CENTERS.

FOLLOWING ARE MAIN PARTICLE INTELLIGENCE SYSTEM CENTERS (PIS) IN PLANTS: PRIMORDIAL CENTRAL PARTICLE INTELLIGENCE SYSTEM CENTERS (C – PIS), PERIPHERAL PARTICLE INTELLIGENCE SYSTEM CENTERS IN PLANTS COMPOSED FROM TWO DIFFERENT INDEPENDENT INTELLIGENCE CENTERS: SENSORY PARTICLE INTELLIGENCE SYSTEM CENTERS (S - PIS) AND MOTOR PARTICLE INTELLIGNCE SYSTEM CENTERS (M - PIS) OF PLANTS.

NANO NEUROLOGY IN PLANTS

THE SENSORY PARTICLE INTELLIGENCE SYSTEMS, MOTOR PARTICLE INTELLIGENCE SYSTEM CENTER AND

CENTRAL PARTICLE INTELLIGENCE SYSTEM CENTER
THE S- PIS, C- PIS, M- PIS ORDERLY HIERARCHY SYSTEMS FUNCTIONS UNDER ONE ARCH –PCS

"PCS - CIRCUITS BETWEEN C- PIS, S- PIS, M-PIS"

THE BIOFRIENDLY VISIBLE OR NON VISIBLE LIGHT FUNDAMENTAL PARTICLES ARE NEEDED AND VITAL FOR MAINTAINING PLANT'S LIFE. THE LIGHT FUNDAMENTAL PARTICLES FROM PLANET SUN EACH DAY TRAVEL AIRBORNE, AT EARTH ENTER INTO INTERNAL ELECTRON NUCLEON SUBSYSTEM UNITS OF THE PLANTS ATOMS IN FLOWERS, FRUITS, LEAVES, STEM, ETC. THE EXOGENOUS LIGHT FUNDAMENTAL PARTICLES (EX. –YFP) ARE USED IN LIGHT PARTICLE COMPOUND CONSTRUCTIONS INTERNAL PLANT CELLS ELECTRONS AND NUCLEON CONSTRUCTIONS AS WELL AS LIGHT PARTICLE STORGAE SYSTEMS INSIDE THE EARTH ELEMENTS IN PARTICLE COMPOUNDS CONSTRUCTION FORMS.

ALL DONE UNDER THE PLANTS – PIS CLOUD ORDERS INSIDE A CLOSED ARCH OR CLOSED CIRCUIT PARTICLE CIRCULATION SYSTEMS, BETWEEN PIS - HIERARCHY CLOUD ORDER SYSTEMS BETWEEN THREE PLANT PIS: STARTING FROM S- PIS REPORTS TO PLANT'S HIGH LEVEL COMMANDEERING C – PIS DOMINANT INTELLECTUAL CENTERS, AND FROM THERE THE C – PIS DECIDED PARTICLE CLOUD ISSUED ORDERS TRAVEL BETWEEN THE C – PIS AND M- PIS PARTICLE CLOUD CIRCULATION SYSTEMS IN CLOSED CIRCUITS BETWEEN THE ABOVE MENTIONED THREE PARTICLE INTELLIGENCE SYSTEM CENTERS CIRCUITS (ONE CIRCUIT – PCS OR ONE ARCH – PCS BETWEEN THREE DIFFERENT PLANT INTELLIGENCE SYSTEM CENTERS OF: S- PIS, C – PIS, M- PIS).

ONE HIERARCHY PCS – CIRCUIT BETWEEN THREE DIFFERENT PIS -CENTERS

THE PLANTS SENSORY PARTICLE INTELLIGENCE SYSTEMS ELECTRONS, NEUTRONS, PROTONS THROUGH STRONG ATTRACTIVE GRAVITON FORCE ORDERS ATTRACT AIRBORNE SUN LIGHT PARTICLES FROM AIR INTO PLANT S – PIS ATOMS, THEREAFTER THE ATTRACTED AND CAPTURED LIGHT FUNDAMENTAL PARTICLES TRANSIT FROM S – PIS VIA PCS (PARTICLE CIRCULATION SYSTEMS) TOWARD DOMINANT HIGHER HIERARCHY C – PIS CENTER FOR THEIR ANALYSIS, DECISION MAKINGS. THE C- PIS EMIT CLOUD – ORDERS ACCORDING TO ENVIRONMENTAL CODITIONS WHAT TO DO AND HOW REACT IN THE BEST INTERESTS OF THE PLANTS ACCORDING EXISTING OUTSIDE ENVIROMENT CONDITIONS. FINALLY, THE ISSURED C- PIS CLOUDS TRANSIT THROUGH PCS BEFORE M- PIS FOR THE EXECUTION OF THE CLOUD ORDERS.

THE S – PIS SENSE AND REPORT ENVIRONMENTAL CLIMATE CONDITIONS THROUGH PCS BEFORE PLANT'S C – PIS DECISIONS. THE C- PIS DECIDE AND EMIT NEEDED PARTICLE CLOUD ORDERS WHICH TRAVEL AND TRANSIT TO M – PIS TO EXECUTE THE CLOUD ORDERS PROPERLY. IN SUNNY DAYS UNDER THE ISSUED CLOUD ORDERS FROM M- PIS CENTERS THE FLOWERS, LEAVES, STEMS, TILT, OR TURN TOWARD THE DIRECTIONS OF SUN WHICH IT IS THE DIRECTIONS OF INCOMING LIGHT FUNDAMENTAL PARTICLES, IN ORDER TO ACCUMULATE AND CAPTURE MAXIMUM QUANTITY DIFFERENT LIGHT FUNDAMENTAL PARTICLES. WHICH ARE VITAL AND NEEDED FOR PLANTS PHYSICAL, CHEMICAL, BIOLOGICAL NEEDS AND FUNCTIONS. ABOVE IS ONE CIRCUIT PARTICLE CLOUD CIRCULATION

SYSTEM ARCH BETWEEN THE THREE DIFFERENT PLANT DOMINANT AND RECESSIVE PARTICLE INTELLIGENT SYSTEM CENTERS (BETWEEN S- PIS, C- PIS, M- PIS).

IN THE EVENTS OF ENVIRONMENTAL HARMFUL WEATHER CONDITIONS, AND EXISTING OUTSIDE HAZARDOUS CONDITIONS, THE S- PIS CLOUD FINDING AND REPORTS, AND C- PIS COMMAND DECISIONS AND CLOUD ORDER, AND M – PIS FUNCTIONS IN REGARDS TO EXECUTION OF THE CLOUDS ORDERS VARY REMARKABLY FROM ONE ENVIRONMENT CASE AND CONDITIONS TO THE OTHER. DEPENDING WHAT IS THE BEST DEFENSIVE PROCEDURES TO SAVE PLANTS LIFE THAT C – PIS DECIDES.

THE PLANT SENSORY SYSTEM CENTER (PIS) MONITOR AND REPORT EXISTANCES OF ENVIRONMENTAL HAZARDOUS CONDITIONS SUCH AS EXISTING BURNING HOT THERMAL PARTICLE CONDITIONS IN MID SUMMER WHICH WILL CAUSE DEHYDRATION AND FIRST OR SECOND DEGREE SUN BURN, OR EXISTANCES OF FREEZING COLD CONDITIONS IN LATE WINTER, OR IN EARLLY SPRING WHICH THE FLOWERS OPEN MAY DIE FROM FREEZING FROST BITES, THE SENSORY PIS SENSE AND REPORT THESE ENVIRONMENTAL INFORMATIONS THROUGH PARTICLE CLOUDS AND TRANSIT TO THE INFORMATION IMAGE CLOUDS BEFORE THE CENTRAL PLANT PARTICLE INTELLIGENCE SYSTEM CENTERS (C – PIS) FOR THEIR INFORMATION, DECISION AND ISSUANCES OF THE ORDER PARTICLE CLOUD ACCORDING EACH GIVEN INDIVIDUAL SITUATIONS.

THE C - PIS IN DAILY SUN SHINE MAY ORDER THE PLANTS FLOWERS, LEAVE STEMS TURN OR TILT TOWARD THE

INCOMING SUN LIGHT DIRECTIONS TO ABSORB MORE LIGHT PARTICLES. OR IN THE CASES OF EXISTING FREEZING COLD IN EARLY SPRING THE C- PIS MAY ORDER THE FLOWER CUSPIDS TO CLOSE AND PROTECT PETALS FROM FROST BITES. DURING SUMMERS BURNING HOT CONDITIONS IN ORDER TO PREVENT FROM FIRST OR SECOND DEGREE BURNS THE C – PIS ORDER THE FLOWERS AND LEAVES TURN DOWN IN THE OPPOSITE DIRECTIONS OF THE INCOMING SUN LIGHTS, IN ORDER TO PREVENT FROM SUN BURNS AND DEHYDRATION.

FOR THE STORAGE OF THE LIGHT FUNDAMENTAL PARTICLES INSIDE ATOMS, THE C – PIS ORDER THE Y- FP TRANSIT INTO INTERNAL ATOM'S SUBSYSTEM UNITS CHEMICAL LAB.S, UNDER A. S. I. FP. Mol. CIC. COMBINE WITH INTERNAL ELECTRONS NUCLEON PARTICLE COMPOUNDS CONSTRUCTIONS OF PLANT FLOWERS, FRUITS, LEAVES, STEM AND PRODUCE LIGHT PARTICLE COMPOUNDS AND CONSTRUCT THE INTERNAL CELL ATOM'S LIGHT FUNDAMENTAL PARTICLE COMPOUND CONSTRUCTIONS FOR STEM, LEAVES, FLOWERS, FRUITS ELECTRONS AND NUCLEONS PARTICLE COMPOUND OR COMPOSITE STRUCTURES.

THESE ARE PHENOMENON OF PLANT ATOM NEO GENESIS, GENESIS OF NEEDED LIGHT PARTICLE COMPOUND CONSTRUCTION NEW ELECTRONS AND NUCLEON CONSTRUCTIONS THROUGH A. S. I. FP. Mol. CIC. IN PARTICLE COMPOUN FORMS. ALL DONE UNDER DIRECT ORDERS OF PARTICLE INTELLIGENCE SYSTEM CENTERS CLOUD ORDERS.

TO ACHIEVE NEO ATOM GENESIS, ELECTRON – NEO GENESIS, NUCLEON GENESIS PHENOMENON UNDER C – PIS, S – PIS, M- PIS CLOUD ORDERS AND DECISIONS.

THE MAIN FUNCTION OF SENSORY PARTICLE INTELLIGENCE SYSTEM CENTERS (S – PIS) IS TO REPORT EXISTING SUN SHINE, OR EXISTING ENVIRONMENTAL HAZARDS SUCH AS FREEZING COLD, OR BURNING HOT SUMMER, ETC. TO THE HIGHER COMMAND CENTRAL PARTICLE INTELLIGENCE SYSTEM CENTERS (C – PIS).

THE ESSENTIAL FUNCTIONS OF PLANTS COMMAND SYSTEMS CENTERS IS TO OPERATE PROTECTIVE AND DEFENSIVE MEASURES, PROVIDE AND MAINTAIN THE CHEMICAL, PHYSICAL, AND BIOLOGICAL NEEDED TASKS, IN NORMAL LEVELS IN THE PLANTS. HELP PLANTS CAPTURE NEEDED LIGHT FUNDAMENTAL PARTICLES, ORDER AND ACHIEVE NORMAL PLANTS FUNCTIONS, AND A. S. I. F.P. Mol. C.I.C., ETC., THE PLANTS C – PIS CENTERS IN DAILY BASIS, ORDER THE PLANTS LEAVES, FLOWERS, STEMS, TO TURN TOWARD INCOMING LIGHT PARTICLE DIRECTIONS, TILT, OR ROTATE, AND TURN TOWARD LIGHT PARTICLE DIRECTIONS WHEN THERE IS SUNSHINE. IN ORDER TO COLLECT LIGHT PARTICLES FROM INCOMING DIRECTIONS OF LIGHT ORIGIN PARTICLES, THROUGH THE FLOWERS AND LEAVES AND STEM ROTATIONS OR TILTS, UNDER THE CIS INTELLECTUAL COMMAND SYSTEMS CENTERS ORDERS THE LEAVES, FLOWERS, STEMS TURN TOWARD LIGHT PARTICLE DIRECTIONS, AND ATTRACT THE LIGHT PARTICLES.

IN DAILY BASIS, THE AIRBORN INCOMING LIGHT PARTICLES DIRECTLY ENTER INTO INSIDE PLANTS INTERNAL ELECTRON NUCLEON SUBSYSTEM –UNITS CHEMICAL LAB.S, AND UNDER A. S. I. F.P. Mol. C.I.C. COMBINE WITH INTER ELECTRON-NUCLEON PARTICLE COMPOUNDS AND CONSTRUCT LIGHT PARTICLE COMPOUNDS OF PLANTS ELECTRON NUCLEON CONSTRUCTIONS, (PHENOMENONS OF PARTICLE COMPOUND NEO- GENESIS AND PHENOMENON OF NEW ELECTRON- NUCLEON –GENESIS) ALL OF THESE NANO-TASKS, SENSED AND REPORTED BY S - PIS AND NECESSARY ORDERS ISSUED THROUGH THE C - PIS AND THE TASKS EXECUTED BY M - PIS CENTERS. ALL PLANTS DEFENSIVE AND PROTECTIVE MEASURES, DAILY PHYSICAL CHEMICAL BIOLOGICAL FUNCTIONS ALL SENSED BY S – PIS, CLOUD ORDERS ISSUED BY C - PIS CENTERS, AND ISSUED ORDERS EXECUTED BY M - PIS CENTERS OF THE PLANTS.

THE S - PIS CENTERS FUNCTIONS ARE SENSING, AND REPORTING, EXISTING ENVIRONMENTAL CONDITIONS, AS WELL AS REPORTING EXISTANCES OF HARMFUL ENVIRONMENTAL CONDITIONS, SUCH AS DEHYDRATIONS, LACK OF IRRIGATIONS ALL GET REPORT TO C - PIS CENTERS FOR THEIR DECISIONS AND ORDERS. AND M – PIS EXECUTE THE ORDERS ACCORDING CLOUD INSTRUCTIONS.

UNDER ABOVE PIS PARTICLE CLOUD ORDERS SYSTEMS (PIS - P. Cl. - OS) EVEN SOME CARNIVOROUS PLANT SPECIES CAN SENSE MOVEMENTS OF INSECTS OVER THE LEAVE SURFACES AND CAN HUNT THE MOVING INSECT AT EXACT LOCATION AND AT EAXCT CORRECT TIME TO STRIKE AGAINST PRAYERS BY LEAVE CLOSURE ENTRAPMENTS. THESE PLANT SPECIES EAT HUNTED LIVE INSECTS SIMILAR

CARNIVOROUS MEAT EATER ANIMALS AS GOOD NUTRIENTS AND MEAL.

PARTICLE CLOUD CURRENTS IN PLANTS
PARTICLE CIRCULATION SYSTEMS (PCS) IN PLANTS

THE PRIMARY P. C. S. CONNECTS THE THREE SENSARY - CENTRAL - MOTOR INTELLIGENCE SYSTEM ELECTRONS NUCLEONS TO EACH OTHER, THROUGH BI-DIRECTIONAL SEPARATE TWO EFFERENT AND AFFERENT PARTICLE CIRCULATION SYSTEM CURRENTS, WHICH THESE TWO OPPOSITE DIRECTION FLOWING CURRENTS, ALWAYS ONE PARTICLE CLOUD CURRENT CIRCULATIONS RUNS IN OPPOSITE DIRECTION OF THE OTHER PARTICLE CLOUD FLOWING CURRENTS.

ALSO SECONDARILY THESE TWO BI DIRECTIONAL RUNING PCS CONNECTS THE ALL PLANTS: (STEMS, FLOWERS, LEAVES, ROOTS, FRUITS) TOTAL EXISTING ELECTRON, NUCLEON, ATOM, CELL POPULATIONS TO EACH OTHER. AND ALL OF THESE BI DIRECTIONAL FLOWING P. C. S. CURRENTS FINALLY CONNECTS THE TOTAL BODY ATOM AND CELL POPULATIONS TO MAIN PRIMARY PLANTS PARTICLE INTELLIGENCE SYSTEMS CENTERS (C – PIS, S – PIS, M – PIS) ELECTRONS –NUCLEONS POPULATIONS THROUGH THESE TWO OPPOSITELY RUNING BI DIRECTIONAL P. C. S.

CURRENTS THE TOTAL PLANTS NANO UNITS, MICRO UNITS, MACRO UNITS PIS CONNECTED TO EACH OTHER, TOTAL BODY NANO UNIT POPULATIONS ARE AWARE FROM FUNCTIONS OF THE EACH OTHER, AND HOW TO COORDINATE AND COOPERATE THEIR TASKS ALTOGETHER.

THROUGH THESE TWO DIFFERENT DIRECTIONAL PRIMARY AND SECONDARY PARTICLE CLOUD CIRCULATIONS SYSTEMS CURRENTS THE ENTIRE PLANTS ELECTRONS NUCLEONS POPULATIONS GET CONNECTION TO EACH OTHER. ABOVE IS THE PLANT'S P.C.S CONSTRUCTIONS BRIEFLY. WHICH THROUGH THESE PCS THE PLANTS INTELLECTUAL CENTERS ARE ABLE TO MANAGE THE OPERATIONS OF THE ENTIRE PLANTS BODY ELECTRONS NUCLEONS PHYSICAL, HEMICAL, BIOLOGICAL FUNCTIONS.

PLANT CLASSIFICATION THROUGH PIS - PCS

UNDER THIS CLASSIFICATIONS THE DIFFERENT PLANTS SPECIES DIVIDE INTO THREE DIFFERENT CLASSES: EITHER PLANTS HAVE ALL THREE SENSARY- CENTRAL-MOTOR PARTICLE INTELLIGENCE SYSTEM CENTERS. THE PLANTS SPECIES WHO DO NOT HAVE ANY KIND INTELLIGENCE SYSTEM CENTERS AT ALL. ALSO THERE ARE THIRD CLASS LARGE NUMBERS OF PLANTS SPECIES WHO POSSESSES INTERMEDIATE AND INCOMPLETE PLANT INTELLIGENCE SYSTEM CENTER. (EXPLAINED IN TEXTS).

THE CLASSIFICATION OF PLANTS UNDER PARTICLE INTELLIGENCE SYSTEMS:

1 - THE PLANT SPECIES WHO POSSESS PIS.

2 - THE PLANTS SPECIES THAT DO NOT HAVE PIS.

3 - THE PLANTS SPECIES WHO POSSESS PARTIAL AND INCOMPLETE: C - PIS, S - PIS, M - PIS.

PLANT'S BRAINS (PIS) FUNCTIONS
THE C – PIS, S – PIS, M – PIS CENTERS PATHO BIOLOGICAL FUNCTIONS

PLANT'S BRAIN'S (PIS) DECISION CAN PREVENT FROM HAZARD

IN LATE WINTER, OR EARLY SPRING, WHEN WARM WEATHER TRIGGER FLOWERING PLANTS, IN CASES OF SUDDEN FREEZING EPISODES, THE S - PIS SENSE ENVIRONMENTAL FREEZING EVENTS, THE S – PIS PRODUCE AND EMIT PARTICLE CLOUD INFORMATION AND IMAGE CLOUDS INTO PLANT'S PCS AND REPORT FREEZING COLD, POSSIBLE FROST BITE, THE S – PIS INFORMATION CLOUDS TRAVEL THROUGH PCS INTO C – PIS.

UNDER CAREFUL ANALYSIS INCOMING INFORMATION IMAGE CLOUDS AND FINDING OF FREEZINF CLIMATE THE C – PIS DECIDE PRODUCE AND EMIT ORDERS FUNDAMENTAL PARTICLE INFORMATION AND IMAGE PARTICLE CLOUDS TO CLOSE FLOWER CUSPIDS, THE C – PIS VIA PCS SEND THE PLATICLE CLOUDS BEFORE M – PIS DECISION AND EXECUTION OF CLOUD ORDERS. THAT THE M – PIS MUST CLOSE FLOWER CUSPIDS, IN ORDER TO PROTECT FLOWER PETALS, OVARIES, STAMEN INSIDE CLOSED CUSPIDS AGAINST FROST BITE AND KEEP THEM SAFE AND PROTECTED INSIDE CLOSED THICK CUSPIDS.

CUSPID UNDER PARTICLE CLOUD ORDERS CLOSE AND PREVENT FROM FROST BITE. IN THE PLANT SPECIES WHO POSSESS INTELLIGENCE SYSTEMS CENTERS, AND C – PIS, S – PIS, M – PIS AS WELL AS PCS THE FLOWERS PROTECTED FROM FROST BITES. M- PIS EXECUTE ORDERS OF C – PIS ACCORDING CLOUDS.

IN MANY PLANT SPECIES WHO DO NOT POSSESS INTELLIGENCE PLANT SYSTEMS CENTERS, THESE SPECIES FLOWERS STAY WIDELY OPEN IN FREEZING COLD, AND SUFFER FROST BITE AND DIE, AND THEY DO NOT PRODUCE ZYGOTTE. BECAUSE THEY COULD NOT SENSE, BECAUSE THEY DO NOT POSSESS INTELLIGENCE SYSTEMS, TO SENSE AND CLOSE CUSPIDS, PROTECT PETALS OVARIES STAMEN FROM FROST BITES.

PLANT'S BRAIN (PIS) OPERATE PLANT'S PHYSIOLOGICAL TASKS

TODAYS EMBRYONIC PHASE TASKS EQUALS TO MILLIONS YEARS AGO EVOLUTION ERA TASKS

THE C – PIS EMITTED PARTICLE CLOUDS (LIVING THING BRAINS EMITTED PARTICLE CLOUDS) CONTROL EMBRYONIC AUTONOMOUS SEQUENTIAL ATOM MUTATIONS AND CELL MUTATIONS, BODY ORGAN GENESIS, WHICH TAKING PLACE OF THESE PROCESSES DURING EMBRYONIC STAGES IS EQUAL TO MOLECULAR EVOLUTION PATHWAYS LAWS AND ORDERS AND MUTATIONS.

IN NORMAL EARLY SPRING CLIMATE, WHEN THERE IS SUN SHINE, DIFFERENT LIGHT FUNDAMENTAL PARTICLES ARE PLENTY AROUND AIRBORNE COMING FROM SUN, UNDER THESE CONDITIONS WHEN THE PLANT'S EMBRYONIC STAGE STEM CELLS MUTATIONS AND DIFFERENTIATING WITH EXPONENTIAL SPEEDS OF A. S. I. FP. Mol. CIC. PLANTS ARE IN EXTREME NEEDS OF ALL KIND DIFFERENT LIGHT FUNDAMENTAL PARTICLES IN ORDER TO CONSTRUCT PLANT'S INTERNAL ELECTRON NUCLEON LIGHT FUNDAMENTAL PARTICLE COMPOUND CONSTRUCTIONS.

AT THIS PHASE PLANTS SENSORY PARTICLE INTELLIGENCE SYSTEM CENTERS (S – PIS) SENSE ENVIRONMENTAL SITUATION, DECIDE, PRODUCE, EMIT INFORMATION IMAGE PARTICLE CLOUDS AND REPORT THE STATUS OF OUSIDE TEMPERATURE, CLIMATE ENVIRONMENT TO THE HIGHER

CENTRAL PARTICLE INTELLIGENCE SYSTEM CENTERS (C- PIS). IN NORMAL FAVORABLE CLIMATES THE PLANT'S C - PIS ISSUE ORDER PARTICLE CLOUDS THAT PLANTS FLOWERS, LEAVES, STEM TO TURN OR ROTATE, TILT TOWARD THE DIRECTIONS OF INCOMING SUN LIGHT FUNDAMENTAL PARTICLES DIRECTIONS AND UNDER PLANT ATOMS, ELECTRONS, NUCLEON'S STRONG GRAVITON FORCES ATTRACT AND ABSORB MAXIMUM QUANTITIES NEEDED LIGHT FUNDAMENTAL PARTICLES FROM AIR.

AIRBORNE LIGHT FUNDAMENTAL PARTICLE ENTER INTO PLANTS, FLOWERS, FRUITS, STEMS, LEAVES INTERNAL ELECTRONS NUCLEONS SUBSYSTEM UNITS CHEMICAL LAB.S COMBINES WITH INTERNAL ELECTRON NUCLEON PARTICLE COMPOUND CONSTRUCTIONS. PRODUCE NEW LIGHT PARTICLE MOLECULAR COMPOUND CONSTRUCTIONS. THESE NEW CONSTRUCTED ATOMS, NUCLEONS, ELECTRONS, CELLS WITH EXOGENOUS LIGHT PARTICLE COMPOUND CONSTRUTCTIONS ALSO ARE STORAGE SYSTEMS OF THE LIGHT PARTICLES.

 AS WELL AS THESE NEWLY CONSTRUCTED ATOMS ELECTRONS, NUCLEONS ARE STORAGE SYSTEMS OF DIFFERENT SUN ORIGIN THERMAL, ELECTRIC, SONIC FUNDAMENTAL PARTICLES AND ALSO STORAGE SYSTEMS OF DIFFERENT ENERGY KINDS, WHICH ARE STORED INSIDE PLANT CELL'S ATOMS, AND IN LATER TIMES CAN BE RETRIEVED AND USED AS NEW SOURCES OF DIFFERENT THERMAL, ELECTRIC, LIGHT, AND SONIC PARTICLES OR ENERGY SOURCES FOR DOING DIFFERENT KINDS NANO TASK AT DIFFERENT INTERNAL ATOM OR INTERNAL PLANT CELL'S QUANTUM LOCATIONS.

ABOVE IN ANIMAL EMBRYONIC PHASES ARE PHENOMENONS OF LIVING THING NEWBORN GENESIS, ATOM GENESIS, CELL GENESIS, ELECTRON NUCLEON GENESIS. WHEN ONE STABLE CELL CHANGE CHEMICAL FORMULARY CONSTRUCTIONS FROM ONE STRUCTURE TO ANOTHER STABLE CHEMICAL FORMULARY STRUCTURE CONSTRUCTION CELLS, ATOMS, ELECTRONS, NUCLAON. ARE IN REALITY NEW AUTONOMOUS SEQUANTIAL STABLE CELL AND ATOM MUTATIONS, NEW STABLE ELECTRON NUCLEON MUTATIONS FROM ONE STABLE CELL, ELECTRON, NEUTRON, PROTON, ATOM TO OTHER STABLE MUTANT CELLS AND ATOMS. (THE CHEMICAL FORMULARY AND CONSTRUCTION CHANGES EQUAL TO THE MUTATIONS).

THESE PRESENT ERA PHENOMENONS IN EMBRYONIC PHASES ARE EQUAL TO CELL DIFFERENTIATION, NEO PLANT & ANIMAL GENESIS, NEO PLANT ORGAN GENESIS, NEO ANIMAL SYSTEM GENESIS OF EMBRYO ARE EQUALS TO PAST ERAS OF THE MOLECULAR EVOLUTION PLANT AND ANIMAL NEO GENESIS PHENOMENONS.

WHICH IN EMBRYONIC STEM CELL DIFFERENTIATION PHASES OF THE PRESENT TIME ERAS, ALL TODAY ORDERED TO BE DONE AND TAKE PLACE UNDER C – PIS, S- PIS, M- PIS ORDER CLOUDS AND PARTICLE INTELLIGENCE SYSTEMS CENTERS DECISIONS TODAYS, WHICH ALL ARE AS SIMILARLY

CHIEVED AND OCCURRED AND CONSTRUCTED ALL DIFFERENT LIVING THINGS SPECIES ONE AFTER THE OTHER IN THE MILLIONS OF YEARS AGO DURING THE PAST ERA OF MOLECULAR EVOLUTIONS.

THE SENSORY SYSTEMS OF S – PIS SENSES OUTSIDE LIGHT PARTICLES STATUS, AS WELL AS THE THERMAL FUNDAMENTAL PARTICLE STATUS AND ELECTRIC FUNDAMENTAL PARTICLE EXISTANCES IN THE AIR AND REPORTS IT'S FINDING BEFORE THE COMMANDING CENTERS OF C- PIS, THAT WHERE ARE THE LOCATIONS, DIRECTIONS OF INCOMING SUN LIGHT PARTICLES. THROUGH PARTICLE CIRCULATION SYSTEMS THE C - PIS RECEIVE ALL INFORMATIONS AND IMAGE CLOUDS, DECIDES, AND IN RESPONSE TO S – PIS PARTICLE CLOUD FINDINGS ISSUE THE COMMAND CENTERS C- PIS ORDER CLOUDS WHICH IS THE ESSENTIAL STEPS WHICH THE PLANT FLOWERS AND STEM AND LEAVES MUST TAKE AND COLLECT MAXIMUM NEEDED QUANTITIES FROM AIRBORNE PARTICLES FOR PLANT'S INTERESTS. FINALLY, MOTOR INTELLIGENCE SYSTEMS CENTERS (M - PIS) EXECUTE CLOUD ORDERS, TURN FLOWERS AND LEAVES TOWARD DIRECTION OF LIGHT PARTICLE TOWARD SUN, TO CAPTURE MAXIMUM NEEDED QUATITIES LIGHT PARTICLES FOR ELECTRON-NUCLEON –GENESIS WHICH ARE SURVIVAL FOR PLANT'S LIFE.

PLANT BRAIN FIGHT AGAINST OUTSIDE HAZARDS TO SAVE PLANTS

IN EXTREME BURNING MID SUMMER TEMPERATURES, THE PLANT'S SENSORY PARTICLE INTELLIGENCE SYSTEMS (S – PIS) SENSES DANGEROUS BURNING, DRYINGS, LETHAL DEHYDRATING OUTSIDE ENVIRONMENTAL CONDITIONS, AND REPORT THROUGH EMISSIONS OF PARTICLE CLOUDS ABOUT THE OUTSIDE HAZARDS INFORMATION AND IMAGE CLOUDS BEFORE THE CENTRAL PARTICLE INTELLIGENCE SYSTEM CENTERS (C – PIS), THROUGH PARTICLE CIRCULATION SYSTEMS (PCS) PARTICLE CURRENTS.

THE COMMANDING CENTRAL PARTICLE INTELLIGENCE SYSTEM CENTERS ELECTRONS, NUCLEONS THINK THANKS ACCORDING RECEIVED CLOUD INFORMATION AND IMAGES DECIDE, PRODUCE AND EMIT PROPER PARTICLE CLOUD ORDERS TO COMBAT AGAINST THE EXISTING OUTSIDE ENVIRONMENTAL CONDITIONS AND EXISTING HAZARDS. THE C- PIS THROUGH EMITTED PARTICLE CLOUDS ORDER PLANT'S MOTOR PARTICLE INTELLIGENCE SYSTEMS CENTERS (M – PIS) TO TURN AND ROTATE AND TILT THE LEAVES, STEMS, FLOWER'S FACE FAR AWAY TO OPPOSITE SUN INCOMING DIRECTIONS, OR ORDERS STEMS AND FLOWERS TO TURN LOOKING DOWNWARD TOWARD THE GROUD, IN ORDER TO DECREASE DIRECT EXPOSURE TO HOT BURNING DESTRUCTIVE SUMMERS POWERFUL THERMAL FUNDAMENTAL PARTICLES OF OUTSIDE WORLD WHICH EASILY CAN CAUSE FIRST AND SECOND DEGREES BURNS AND KILL THE PLANT IN SPOT. PREVENT AND AVOID FROM DIRECT BURNING LIGHT PARTICLES EXPOSURES.

IN SPECIES WHO DO NOT POSSESS PLANT INTELLIGENCE SYSTEM CENTERS, THESE SPECIES GET HARMED FROM HEAT STROKE, BURNS, DEHYDRATIONS, MUCH EASIER THAN THOSE WHO HAVE PRIMARY PARTICLE INTELLIGENCE SYSTEM CENTERS.

PHENOMENON OF PLANT IMMUNITY AND DEFENSIVE SYSTEMS, PLANT'S ALZHEIRMER DISEASE

PLANT'S AGING EFFECTS PLANTS IMMUNITY AND DEFENSE SYSTEMS ALSO CAUSE DETERIORATION AND SLOWING DOWN IN PLANTS INTELLIGENCE SYSTEMS CENTERS.

THE YOUNG HEALTHY FLOWERS AND PLANTS POSSESSES MUCH SHARP AND BETTER IMMEDIATE DEFENSE SYSTEMS DELIVERY AND IMMUNITY OPERATIONS AND THEIR SENSARY SYSTEMS, CENTRAL INTELLIGENNCE SYSTEMS, MOTOR SYSTEMS ALSO FUNCTION SHARPER AND MUCH BETTER THAN OLDER SLOW AND SICK LOOKING PLANTS, WHICH THEIR HEALTH CONDITIONS MOSTLY SHOULD BE CALLED AS BORDER LINE.

IN ADDITION, IN SOME SPECIES THE IMMUNITY SYSTEMS, DEFENSE SYSTEMS, PARTICLE INTELLIGENCE SYSTEM CENTERS ARE INHERITENTLY MUCH BETTER THAN THE OTHER SPECIES WHO DO NOT POSSESSED OR NOT

DEVELOPED THESE SYSTEMS. ADDITIONALLY, IN OLDER AGE PLANTS EVEN THE CENTRAL INTELLIGENCE SYSTEMS OF PLANTS SLOWS DOWN AND FUNCTIONS CAN NOT BE COMPARED TO YOUNG AGE PLANTS AT SAME KIND SPECIES AT ALL.

IN EARLY SPRING WHEN OLD AND YOUNG AGE PLANTS START FLOWERING, THE PROTECTIVE MOVEMENTS OF FLOWERS AND LEAVES, STEMS AND INTELLIGENCE SYSTEMS FUNCTIONS, IN COMPARING THE OLD PLANTS, WITH YOUNG PLANTS PERFORMANCES IN THE SAME SPECIES ARE DIFFERENT, IN OLDER PLANTS IT DETERIORATED AND CHANGED IN VARIABLE DEGREES AT OLDER PLANTS. MOST FUNCTIONS IN OLDER PLANTS BECOME MUCH SLOW AND WEAKER FUNCTIONS THAN THE YOUNGER PLANTS. THE YOUNG PLANTS PERFORM EVERY FUNCTIONS ACCURATE, SHARP, AND FAST, AND OLDER PLANTS IT LOOKS CLEARLY OLDER PLANTS EVEN MAY FORGET TO DO THEIR JOBS IN MANY FUNCTIONS AS THEY WERE DOING IN YOUNG AGES, THIS IS A KIND OF PLANT'S ALZHEIMER DISEASE.

WHEN COMPARING THE SAME SPECIES YOUNG AND OLD PLANTS NEXT TO EACH OTHER IN A FREEZING OR SNOWING EARLY SPRING MORNING, THE GENERAL PROTECTIVE FUNCTIONS OF YOUNG AGE PLANTS ARE FAR BETTER, THE YOUNG PLANTS START CLOSING CUSPIDS WHEN CLIMATE CHANGES TO FREEZING. BUT THE OLDER PLANTS EVEN HAVE NOT STARTED TO DO THE SAME, EVEN THEIR AGED INTELLIGENCE SYSTEM CENTERS ARE DETERIORATING, FORGETING, SLOWING DOWN AND NOT FUNCTIONING EVEN MAY NOT DO FUNCTION AT ALL, WHAT THE YOUNG

PLANTS ARE DOING. THE PLANT ALZHEIMER IS SERIOUS THAN ANIMAL SPECIES.

EVEN IN SAME AGE PLANTS ALSO ARE A LOT OF DIFFERENT TYPES VARIABLES, THE IMMUNITY SYSTEMS OF SAME SPECIES PLANTS WITH SAME AGES IN ONE PLANT SPECIES COMPARING TO OTHER, THE FUNCTIONS ALSO ARE VARIED, ONE PLANT SPECIES DO BETTER THAN THE OTHER.

THE PLANTS WITH NO PARTICLE INTELLIGENCE SYSTEM

THE PLANTS PARTICLE INTELLIGENCE SYSTEMS CONTROL STEM CELLS RAPID GROWTH, SEQUENTIAL STABLE MUTATIONS (CELL DIFFERENTIATION) IN EARLY SPRING WHICH SEQUENTIAL CELL MUTATIONS WITH RAPID CELL MULTIPLICATIONS AND GROWTH TAKE PLACE WITH EXPONENTIAL SPEEDS, ALSO PIS CONTROL ENTIRE PHYSICAL CHEMICAL AND BIOLOGICAL FUNCTIONS OF PLANTS.

THERE ARE A LOT OF PLANT SPECIES, WHO DO NOT HAVE INTELLIGENCE SYSTEM CENTERS? AND HAVE NOT DEVELOPED FUNCTIONING PLANT INTELLIGENCE SYSTEMS. THEREFORE, IN THESE SPECIES THE DESCRIPTION OF PLANT FUNCTIONS UNDER C- PIS, S – PIS, M - PIS ARE NOT EXISTED IN DETECTABLE LEVELS AT ALL. OR THESE PIS FUNCTIONS ARE AT QUANTUM LVELS, NOT SENSABLE LEVELS AND NOT NOTICEABLE PRESENTLY. ADDITIONALLY, THERE ARE MANY PLANT SPECIES ONLY HAVE PARTIALLY DEVELOPED

INTELLIGENCE SYSTEMS, AND INCOMPLETE INTELLIGENCE SYSTEM CENTERS, THESE SPECIES RESPONSES ARE NOT GOINIG TO BE THE SITUATIONS DESCRIBED ABOVE.

IN THOSE SPECIES WITH EXISTING C – PIS, S – PIS, M - PIS IN EARLY SPRING DURING FAST- GROWTH, STEM CELL'S PHASE, WHEN THE NEWLY BORN EMBRYONIC STEM CELLS START THEIR EXPONENTIAL SPEED MULTIPLICATIONS AND DUPLICATION FAST EXPONENTIAL GROWTH PHASES, ALL EMBRYONIC PHASES TAKE PLACE UNDER SENSARY REPORTS OF S - PIS AND UNDER DIRECT PARTICLE CLOUD ORDERS OF C – PIS, AND ORDER EXECUTIONS DONE UNDER M - PIS ACCORDINGLY. AND ALL THREE PIS CENTERS MUST COORDINATE THEIR FUNCTIONS ACCORDING CELLS AND ATOM'S PARTICLE INTELLIGENCE SYSTEMS CENTERS ORDERS.

AT EARLY STAGE EMBRYONIC PLANT STEM CELL GROWTH PHASES WHEN THE FLOWERS AND BUDS OPENS DURING THE DAY, FOR PROTECTION OF EARLY STAGE PETALS, OVARIES, STAMEN IN COLD CHILLING NIGHTS, AND EARLY MORNING HOURS, UNDER PARTICLE CENTRAL INTELLIGENCE SYSTEM CENTERS ORDER, WHEN THE WEATHER SENSED AS FREEZING LEVELS, THE CUSPIDS CLOSE AT NIGHTS AND EARLY STAGE PETALS, STAMEN AND OVARIES GET PROTECTION FROM CHILLING COLDS DAMAGES OF NIGHT AND EARLY MORNING TIMES, UNDER CUSPID CLOSURE ORDERS WHICH GET ACHIEVED WITH COOPERATIONS OF ALL C – PIS, S- PIS AND M – PIS CENTERS ALTOGETHER COORDINATED WITH EACH OTHER.

AND SHORTLY THEREAFTER THE FREEZING NIGHT TIMES, WHEN AT THE DAY ARRIVES WITH SUN RISE AND WARMER CLIMATE, IMMEDIATELY THE S – PIS SENSE THE THERMAL PARTICLES IN AIR WITH AIRBORNE ALL KIND LIGHT PARTICLES AROUND, THE C – PIS AFTER GETTING THE REPORTS FROM S – PIS CENTERS REPORTS. AT THIS TIME THE C – PIS ISSUE AND EMIT PARTICLE CLOUD ORDERS FOR OPENING FLOWER CUSPIDS AND BUDS STARTS TO OPEN IN DAY TIMES AGAIN, DURING THE SUN SHINE THE FLOWERS CUSPIDS STAY WIDELY OPENS ALL DAY LONG.

EVEN UNDER SEPARATE ANOTHER SET PARTICLE CLOUD REPORTS BY S – PIS, AND DIFFERENT CLOUD ORDERS BY C – PIS THE PLANTS STEMS, FLOWERS, LEAVES, HAVE STARTING TO TILTS, BENDS AND ROTATE TOWARD THE SUN DIRECTIONS, AND CHASE THE DIRECTION OF SUNS TRAVEL PATH FROM EAST TO WEST EVERY DAY. UNDER WELL FUNCTIONING PLANTS CENTRAL INTELLIGENCE SYSTEMS CENTERS ORDER. DURING THESE PROCEDURES THE PLANTS STEMS, FLOWERS, LEAVES ACCUMULATING NEEDED LIGHT FUNDAMENTAL PARTICLES, IN ORDER THOSE LIGHT PARTICLES TO BE USED INSIDE THE SUBSYSTEM CHEMICAL LAB.S OF DIFFERENT ELECTRONS, NUCLEONS UNDER A. S. I. FP. Mol. CIC. FOR CONSTRUCTIONS OF NEW LIGHT FUNDAMENTAL PARTICLE COMPOUNDS, NEO GENESIS OF NEW PLANT ATOMS AND NEW PLANT CELLS, AS WELL AS NEO LIVING THING GENESIS.

AT THE SAME TIME, THE FLOWERS WHOSE ELECTRONS, PROTONS, NEUTRONS AND ATOMS CONSTRUCTIONS WHICH THEIR MOLECULAR CONSTRUCTIONS MOSTLY HAVE BEEN USED THROUGHELECTRIC FUNDAMENTAL PARTICLE

MOLECULAR COMPOUND CONSTRUCTIONS. THESE FLOWERS AND PLANTS WITH ELECTRIC PARTICLE MOLECULAR CONSTRUCTIONS POSSESSES MUCH HIGHER LEVEL IMMUNITY AND DEFENSIVE SYSTEMS THAN ORDINARY LIGHT FUNDAMENTAL PARTICLE CONSTRUCTED ATOMS, AND CELLS. THESE STRUCTURES PROVIDE BETTER IMMUNITY AND DEFENSE SYSTEM TO PLANTS

EFFECT OF ATOM PARTICLE CONSTRUCTIONS ON PLANT IMMUNITY AND DEFENSE SYSTEMS

==

THE PLANT'S FLOWERS, LEAVES, STEMS THAT THEIR PARTICLE ELECTRONS NUCLEONS FUNDAMENTAL PARTICLE MOLECULAR COMPOUND CONSTRUCTIONS HAVE BEEN MADE FROM USES OF HIGH ENERGY BIOFRIENDLY PINK COLOR ELECTRIC PARTICLE COMPOUND CONSTRUCTIONS, HAVE STRONGER AND BETTER PLANT IMMUNITY AND DEFENSIVE SYSTEMS THAN OTHER THE OTHER COLOR PLANTS WHICH HAVE BEEN MADE FROM OTHER COLOR PARTICLE COMPOUND CONSTRUCTIONS.

THE PLANT CELLS, ATOMS WHOSE FUNDAMENTAL PARTICLE MOLECULAR COMPOUNDS ARE MADE FROM VIOLET COLOR BIOFRIENDLY ELECTRIC FUNDAMENTAL PARTICLE COMPOUND CONSTRUTIONS HAVE BETTER IMMUNITY SYSTEMS AND BETTER DEFENSIVE SYSTEMS AGAINST HARSH OUTSIDE ENVIRONMENTAL CONDITIONS (SUCH AS EXTREME FREEZING COLD IN EARLY SPRING, OR HARSH HOT CLIMATE CONDITIONS IN MID SUMMER), THESE PLANTS HAVE BETTER SIVIVAL RATE THAN THE OTHER COLOR FUNDAMENTAL PARTICLE COMPOUND CONSTRUCTED PLANTS, FLOWERS, LEAVES AND STEMS.

THE VIOLET COLOR FUNDAMENTAL PARTICLE ARE BIOFRIENDLY ELECTRIC – LIGHT PARTICLES FROM SUN ORIGIN WITH MORE ELECTRIC ENERGY CONTENT PER EACH ONE FUNDAMENTAL PARTICLE CONTENT.

THE PINK OR VIOLET COLOR FLOWERS, STEMS, LEAVES WHICH HAVE BEEN CONSTRUCTED FROM VIOLET PARTICLE COMPOUNDS CELL, ELECTRONS, NUCLEONS, ATOMS COMPOUND CONSTRUCTIONS, THEIR IMMUNITY AND DEFENSIVE SYSTEMS ARE BETTER THAN OTHER COLOR FUNDAMENTAL PARTICLES TO SURVIVE IN HARSH CLIMATE CONDITIONS.

THE FLOWERS THAT HAS BEEN CONSTRUCTED FROM WHITE COLOR LIGHT PARTICLE –COMPOUNDS OR HAVE BEEN CONSTRUCTED FROM YELLOW COLOR LIGHT PARTICLE COMPOUNDS POSSESS LESS ELECTRIC ENERGY CONTENTS, AND THEIR DEFENSIVE SYSTEMS AND IMMUNITY ARE LESS THAN THE FLOWERS WHO HAVE BEEN CONSTRUCTED

FROM VIOLET COLOR LIGHT PARTICLE COMPOUND CONSTRUCTIONS, OR VIOLET COLOR FLOWERS.

VIOLET COLOR FLOWER SPECIES IN HARSH DANGEROUS ENVIRONMENTAL CONDITIONS SUCH AS EXTREME COLD OR EXTREME HOT WEATHERS HAVE BETTER CHANCE OF SURVIVAL THAN THE WHITE OR YELLOW COLOR FLOWERS, HAVING MORE ELECTRIC ENERGY CONTENT VIOLET COLOR PARTICLE COMPOUNDS PROVIDE BETTER RESISTENCES AND IMMUNITY AND DEFENSE SYSTEMS AGAINST HOT WEATHER AS WELL AS AGAINST COLD WEATHER HAZARDOUS CONDITIONS, WHICH UNDER THOSE LEVEL HEAT OR FREEZING CONDITIONS MOST WHITE OR YELLOW FLOWERS DIE, BUT THE VIOLET FLOWERS AND PLANTS MAY SURVIVE.

PLACING VIOLET COLOR AND WHITE COLOR PLANTS UNDER COLD ENVIRONMENTS, THE VIOLET COLOR FLOWERS SURVIVE BETTER THAN WHITE COLOR FLOWERS, IT IS THE SAME, PLACING VIOLET AND WHITE COLOR FLOWERS UNDER EXTREME HARSH HOT SUMMER TEMPERATURES WITH DEHYDRATIONS, THE VIOLET COLOR PLANTS POSSESS BETER CHANCES OF SURVIVAL THAN THE WHITE COLOR FLOWERS.

HIERARCHY ORDER SYSTEMS BETWEEN

S- PIS, C – PIS, M - PIS IN PLANTS

THE PLANTS, FLOWERS, LEAVES, STEMS AND ROOTS SENSORY FUNDAMENTAL PARTICLE INTELLIGENCE SYSTEM CENTERS (S - PIS) SENSE OUTSIDE WORLD'S ENVIRONMENTAL IMAGE AND INFORMATION CONDITIONS, DO RESEARCH WORKS, THINK, PRODUCE SENSORY DECISION RESULT PARTICLE CLOUDS ABOUT OUTSIDE ENVIRONMENT CONDITIONS AND EMIT THE LIGHT, THERMAL, ELECTRIC, SONIC FUNDAMENTAL PARTICLE INFORMATION AND IMAGE PARTICLE CLOUDS Y. T. E. S. – FP – I.I. – P. Cl.) INTO PLANT'S GENERAL PARTICLE CIRCULATION SYSTEMS (PCS).

THE S - PIS REPORT THROUGH PARTICLE CLOUDS THE EXISTING OUTSIDE WORLD ENVIRONMENTAL IMAGES AND INFORMATIONS SUCH AS EXISTENCES OF AIRBORNE LIGHT, THERMAL, ELECTRIC, SONIC FUNDAMENTAL PARTICLES CLOUDS (Y. T. E. S. – FP – I.I. – P. Cl.) AND DIRECTION OF THE INCOMING PARTICLES, THE EXISTING OUTSIDE TEMPERATURE, COLD WEATHER, OR WARM MODERATE BIOFRIENDLY LIGHT, ELECTRIC, THERMAL, SONIC FUNDAMENTAL PARTICLES, OR EXITENCES OF SEVERE HOT BIOLOGICAL HOSTILE AIRBORNE THERMAL FUNDAMENTAL PARTICLES AND THEIR DIRECTIONS, AT THE SAME PATTERNS THE S – PIS OF ROOTS SENSE, ANALIZES, PRODUCE PARTICLE CLOUDS AND EMIT INFORMATION AND IMAGE PARTICLE CLOUDS ABOUT EXISTING WATER, DIFFERENT MINERALS,

DRY SOIL, OR ANY OTHER BIOFRIENDLY OR BIOHOSTILE EXISTING ENVIRONMENTAL CONDITIONS, EMIT INFORMATION AND IMAGE CLOUDS INTO PCS, TO BE CARRIED BEFORE C- PIS DECISION AND HIERARCHY COMMAND PARTICLE CLOUD ORDERS, TO BE CARRIED OUT BY C – PIS AND CONSEQUENT ORDER CLOUDS BY M – PIS DONE FOR BENEFIT OF THE PLANTS, ROOTS, STEMS, FLOWERS, FRUITS, LEAVES, ETC.

PLANT'S: S – PIS, C – PIS, M – PIS ALWAYS ARE IN CLOSE CONTINUOUS CONSULTS AND CONTACTS THROUGH T. S. Y. E.- FP – I.I. – P. Cl. AND CONTROL ENTIRE PLANT'S BIOLOGICAL, PHYSICAL, CHEMICAL FUNCTIONS OF ENTIRE PLANT'S ELECTRONS, PROTONS AND NEUTRONS POPULATIONS FUNCTIONS WITH FULL COORDINATIONS AND COOPERATIONS OF THE EACH OTHER.

THE INCOMING Y. S. T. E. - FP- I.I.- P. Cl. FROM SENSORY PARTICLE INTELLIGENCE SYSTEM CENTERS, UPON ARRIVAL INTO PLANT'S FUNDAMENTAL PARTICLE INTELLIGENCE SYSTEM (C – PIS) CENTER'S IN THE PLANT'S CELLS AND ATOMS THROUGH GENERAL PARTICLE CIRCULATION SYSTEMS (PCS) ENTER INTO INTERNAL ELECTRON - NUCLEON SUBSYSTEM – UNITS CHEMICAL LAB.S, COMBINE WITH INTERNAL ELECTRON- NUCLEON PARTICLE COMPOUNDS, PRODUCE S. T. Y. E. – FP – I.I. – P. Cl. – COMPOUND MOLECULAR CONSTRUCTIONS OF THE PLANTS, LEAVES, FLOWERS, STEMS, ROOTS FUNDAMENTAL PARTICLE CLOUD COMPOUND INFORMATION AND IMAGE CONSTRUCTIONS, THIS PHENOMENON ALSO IS STORAGE SYSTEMS OF INFORMATIONS AND IMAGES INSIDE THE ELECTRONS, NUCLEONS IN PLANTS.

THIS PHENOMENON IS KNOWLEDGE – GENESIS PROCESS, PLANT'S PSYCHE GENESIS PHENOMENON AS WELL, WHICH ARE COLLECTING OUTSIDE INFORMATIONS AND IMAGE RECORDS AT INSIDE THE PLANTS INTERNAL ELECTRONS AND NUCLEONS PARTICLE CLOUD COMPOUND CONSTRUCTIONS AS A REFERENCE KNOWLEDGE AND INFORMATION SOURCES EXISTING ENCYCLOPEDIA INSIDE THE PLANTS BRAINS CELLS AND ATOMS READY FOR USE AT ANY GIVEN TIME OF THE PLANTS LIFE. WHICH PLANT BRAIN FOR DECISION MAKINGS COMPARE NEWLY ARRIVING INCOMING PARTICLE CLOUDS INFORMATIONS WITH ALL RELATED PRE EXISTING STORED INFORMATION IMAGE PARTICLE CLOUDS.

THE INDIVIDUALLY CREATED PLANT'S DIFFERENT: S – PIS, C- PIS, M- PIS, AT MILLIONS YEARS AGO DURING THE MOLECULAR EVOLUTION USED THE EXACT SIMILAR PATTERNS OF INTELLIGENCE SYSTEMS CENTERS FUNCTIONS PROCESS AND CREATED TO DAYS ENTIRE EXISTING THINGS OF UNIVERSE WHICH WE SEE AT PRESENT TIME.

THE TODAYS, LIVING THING BRAIN FUNCTIONS ARE EXACTLY THE SAME CREATED COPY PATTERNS WHICH WAS CREATED, OCCURRED AND STARTED TO FUNCTION AND CREATE MORE CREATIONS ONE AFTER THE OTHER IN MILLIONS YEARS AGO DURING THE EVOLUTION. AND TODAYS LIVING THING PSYCHE FUNCTIONS EQUALS TO THE EVOLUTION ERA BRAIN FUNCTIONS COPIES, WHICH ARE TAKING PLACE AUTONOMOUSLY AND SEQUENTIALLY SINCE THEN. THE PLANTS BRAINS DECISION MAKINGS UNDER A. S. I. FP. Mol. CIC. AND DIFFERENT SYSTEMS OF USING INFORMATION IMAGE PARTICLE CLOUDS CURRENTS TODAYS ARE BIOLOGICALLY EQUAL TO THOUGHT CURRENT,

PIS - PSYCHE AND KNOWLEDGE OPERATIONS AND CREATIONS OF PAST EVOLUTION ERA PROCESSES EXACTLY SAME.

THE DECISION MAKINGS OF PLANTS DURING ENCOUNTERING WITH COLD, HOT, OR NORMAL, OR ABNORMAL CONDITIONS ALL COMPARED WITH BRAINS STORED KNOWLEDGE SYSTEMS INFORMATIONS AND CLOUDS ENCYCLOPEDIA WITH STANDARD NORMS, THEREAFTER THE C- PIS CENTERS THINK AND DECIDE OVER THOSE STORED INFORMATION AND IMAGES OF NORMAL AND STANDARDS ONES. THEREAFTER THE LIVING THINGS DECIDE TO ORDER HOW THE PLANT STEMS, LEAVES, FLOWERS MUST TILT OR ROTATE IN WHICH DIRECTIONS AND PATTERNS ACCORDING THE PARTICLE CLOUD DECISIONS AND ORDERS THROUGH C – PIS, C- PIS, AND M – PIS THE FLOWERS, STEMS, LEAVES CHANGE FUNCTIONS FROM ONE SITUATIONS TO THE OTHER.

PSYCHE GENESIS AND THOUGHT CURRENT GENESIS IN PLANTS

KNOWLEDGE ACCUMULATION INSIDE PLANTS ATOMS AND CELLS

PLANT'S DECISION MAKINGS

LIVING THINGS STORE LIGHT, THERMAL, SONIC AND ELECTRIC FUNDAMENTAL PARTICLE INFORMATION AND IMAGE PARTICLE CLOUD COMPOUND CONSTRUCTIONS (Y. S. T. E. – FP – I.I. – P. Cl. – COMP.) INSIDE THE PLANTS, FLOWERS, ROOTS, STEMS INTERNAL ELECTRON NUCLEON CONSTRUCTIONS. THESE INFORMATION AND IMAGE PARTICLE CLOUD COMPOUND CONSTRUCTIONS INSIDE PLANT ATOMS AND CELLS ARE STORAGE AND CONSTRUCTION OF KNOWLEDGE, INFORMATIONS, IMAGES OF OUTSIDE SUBJECTS IN PARTICLE CLOUD MOLECULAR FORMS INSIDE PLANTS ATOMS.

WHICH THE CLOUDS ARE EXACT COPIES OF SUBJECTS MADE FROM PARTICLE CLOUDS IN SUBJECT- FORMS. THESE STORED PARTICLE CLOUD INFORMATIONS ARE SIMILAR TO ENCYCLOPEDIA INFORMATION AND IMAGE RECORDED AND KEPT INSIDE A DISC SYSTEM.

THE PLANTS CENTRAL PARICLE INTELLIGENCE SYSTEM CENTERS (C – PIS) OF ATOMS, ELECTRONS, AND NUCLEONS HAVE STORAGE OF THE EXISTING OUTSIDE SUBJECTS INFORMATIONS AND IMAGES, WHICH THEIR INTERACTIONS UNDER A. S. I. FP. Mol. CIC. BIOLOGICALLY SENSE AS PSYCHE (SOUL), AND THOUGHT CURRENTS OF LIVING THINGS.

THIS IS PHENOMENON OF PARTICLE CLOUD COMPOUND GENESIS (CONSTRUCTION) INSIDE PLANT'S BRAINS ATOMS, ELECTRONS, NUCLEONS. THESE ARE THE CLOUD MOLECULES THAT CONSTRUCT THE PLANT – BRAIN'S ATOMS, CELLS CONSTRUCTIONS. THESE CLOUD MOLECULES ARE STORAGE SYSTEMS OF THE OUTSIDE SUBJECTS INFOMATIONS AND IMAGES IN PARTICLE CLOUD FORMS AT

INSIDE THE PLANTS INTERNAL ELECTRONS AND NUCLEONS CONSTRUCTIONS (AT PARTICLE CLOUD COMPOUND FORMS).

THOUGHT CURRENT AND PSYCHE GENESIS IN LIVING THINGS

THE C - PIS THROUGH A. S. I. F.P. Mol. C.I.C. COMPARE AND ANALIZE INCOMING PARTICLE CLOUDS INFORMATIONS AND IMAGES FROM S – PIS PARTICLE CLOUD ORIGINS. THE C – PIS COMPARE NEW INCOMING PARTICLE CLOUDS (FROM S – PIS) WITH ENTIRE PREVIOUSLY EXISTING INTERNAL ELECTRON NUCLEON PARTICLE COMPOUND INFORMATIONS AND IMAGES. AND THE C - PIS THEREAFTER THINK, DECIDE AND MAKE PROPER FINDANL DECISIONS OVER THESE INTERNAL ELECTRON NUCLEONS INFORMATIONS AND IMAGE COMPARISSIONS WITH EACH OTHER THROUGH EXPONENTIAL INTERACTIONS OF MILLIONS OF SEQUENCES SPEEDS FOR A QUANTUM TIME UNIT. THEREAFTER THE C - PIS ISSUE EMISSION OF PARTICLE CLOUD ORDERS TO BE TRANSMITTED TO LOWER HIERARCHY SYSTEMS, BEFORE THE MOTOR INTELLECTUAL SYSTEMS CENTERS (M – PIS), THE ORDERS TO BE EXECUTED ACCORDINGLY AS ISSUED BY C - PIS EXACTLY. UNDER HIERARCHY SYSTEMS FROM HIGHER CENTERS TO LOWER INTELLIGENCE SYSTEM CENTERS, THE M - PIS EXECUTE ORDERS AS RECEIVED FROM C – PIS EXACTLY ACCORDING ORDERS INSTRUCTIONS. IN COLD FREEZING WEATHER THE M- PIS CLOUDS ORDERS CLOSES FLOWER CUSPIDS, IN

NORMAL SUNSHINE SPRING CONDITIONS THE M- PIS ROTATE FLOWERS TOWARD THE SUN AND DIRECTION OF THE INCOMING SUN LIGHTS, AND IN HOT SUMMERS IN DEHYDRATING THERMAL PARTICLES EXPOSURES M- PIS ORDERS TURN DOWN FLOWERS FACE AWAY FROM DIRECT SUN EXPOSURES, ETC.

FUNDAMENTAL PARTICLE WEAPONS OF MASS DESTRUCTION, AND CRIMES AGAINST HUMANITY

SUPER-SONIC PARTICLE WEAPONS INJURIES
MICROWAVE WEAPONS OF MASS DESTRUCTION AND MICROWAVE WEAPON WOUNDS,

THE CRIMES AGAINST HUMANITY

SHOOTING MICROWAVE SUPERSONIC LIGHT CODED LASER PARTICLES WEAPONS OF MASS DESTRUCTION CAN BE FIRED AIRBORN FROM ANY DISTANCE, FROM ARMY BASE OR FROM POWERFUL TELECOMMUNICATION TOWERS,

THROUGH COMPUTERIZED WELL CALIBRATED AND DOSE ADJUSTED VARIABLE LETHAL AND NON-LETHAL MICROWAVE DOSES DELIVERED COMPUTERIZED INTO ANY SELECTED HOUSES IN ANY CITY. THE RULERS AGENTS CAN SHOOT OR TORTURE BY MICROWAVE SHOTS ANYONE INSIDE THEIR OWN HOMES WITHOUT LEAVING ANY TRACE, USE OF THESE DEVICES DO NOT LEAVE TRACE OF INJURIES FROM SHOTS AND KILLING IN BEHIND.

MICROWAVE SUPERSONIC WEAPONS INJURIES ARE INTERNAL ELECTRON-NUCLEON PARTICLE COMPOUND DESTRUCTIONS, THE PARA- CLINICAL STUDIES SUCH AS CHEMICAL STUDIES, MRI, CAT SCANS, PET SCAN OR SIMILAR STUDIES NOT HELPFUL AT MOST, THESE ABUSES ARE COMMON IN LESS ADVANCED REGIONS OF THE WORLD NATIONS. CONTRARY TO PARA CLINICAL MANIFESTATIONS, THE INTERNAL ATOM INJURIES BY LASER MICROWAVE PRODUCE PATIENT'S COMPLAINTS, SIMILAR TO COMPLAINTS OF INJURIES CAUSED BY OTHER BODY INJURIES AND DISEASES.

DESTRUCTION OF INTERNAL ELECTRON NUCLEON S. FP - I.I. - P.cl. COMP. CONSTRUCTION BY MICROWAVE

PATHOLOGY OF MICROWAVE LASER INJURIES

Head Injuries

POWERFUL DESTRUCTIVE BIOCIDAL SUPERSONIC MICROWAVE FUNDAMENTAL PARTICLES SHOTS FROM LASER WEAPONS OF MASS DESTRUCTION AIRBORNE ENTER

INTO INSIDE CNS AUDITARY CENTERS INTERNAL ELECTRONS- NUCLEONS SUBSYSTEM UNITS CHEMICAL LAB.S AND DESTRUCT FINE S. Y. – F.P. – I.I. – P. Cl. - COMPOUNDS CONSTRUCTIONS OF THE CNS ELECTRONS –NUCLEONS MOLECULAR STRUCTURES. MOST OF THESE PARTICLE- CLOUD COMPOUND CONSTRUCTIONS ARE IMPORTANT FACTORS AT PRODUCTION OF THOUGHT CURRENTS AND HUMAN PSYCHE, HUMAN KNOWLEDGE AND INTELLIGENCE SYSTEMS.

IN THE CNS THE S. Y. T. E. – F.P. I.I. – P. Cl. CURRENTS INTERACTIONS WITH EACH OTHER BETWEEN THE DIFFERENT CNS CENTERS ELECTRONS AND NUCLEONS UNDER A. S. I. FP. Mol. CIC. PRODUCE PSYCHE AND THOUGHT CURRENT, LASER MICROWAVE DESTRUCTIVE POWERFUL PARTICLE CURRENTS CAN DESTROY THESE SYSTEMS WITHOUT LEAVING ANY DETECTIBLE TRACES. MICROWAVE LASER SHOTS DESTROY CONSTRUCTIONS OF THE BIOFRIENDLY PRE-EXISTING PREVIOUSLY CONSTRUCTED INTERNAL CNS ELECTRONS AND NUCLEONS E. T. Y. S. –F.P. – I.I. – P. Cl. - COMPOUND CONSTRUCTIONS.

IN CONTINUATION OF CNS ELECTRON-NUCLEON PARTICLE COMPOUND DESTRUCTIONS, THE HARSH BIO-CIDAL MICROWAVE PARTICLES COMBINE WITH REMNANTS OF INTER ELECTRON- NUCLEON PARTICLE COMPOUNDS UNDER REGENERATIVE OR DEGENERATIVE A. S. I. F.P. Mol. C.I.C. AND PRODUCE NEWLY CONSTRUCTED NON-LIVE MICROWAVE PARTICLE-COMPOUNDS CONSTRUCTIONS INSIDE CNS AUDITARY CENTERS ELECTRON-NUCLEONS,

THE PRODUCED ROARING MICROWAVE PARTICLE COMPOUND CONSTRUCTIONS INSIDE CNS ELECTRONS – NUCLEONS CAUSE CREATIONS OF PERSISTENT HARSH ROARING TINITUS IN PATIENTS HEAD WHICH THESE TINITUS ARE NON INTERRUPTIVE AND PERMANET TINITUS WITH NO RELIEF, AND RELATE TO EXISTENCES OF MICROWAVE PARTICLES COMPOUNDS IN CNS ELECTRONS-NUCLEONS PARTICLE COMPOUND CONSTRUCTIONS'

ABOVE IS PROCESS AND CREATION OF NON LIVE BIOCIDAL MICROWAVE PARTICLE COMPOUND MOLECULES.

THIS IS NEO-GENESIS OF PATHOLOGICAL CONSTRUCTED ATOMS PHENOMENON INSIDE CNS ELECTRONS NUCLEONS, WHICH THESE POTHOLOGICALCONSTRUCTED ELECTRONS AND NUCLEONS ARE REPLACING THE PREVIOUSLY NORMAL CONSTRUCTIONS OF THE CNS ATOMS. (PHENOMENON OF NEO GENESIS OF PATHOLOGICAL ATOM CONSTRUCTIONS).

PHYSIOLOGICAL FUNCTIONS OF CERUMEN

WHEN PATIENT'S HEAD SHOT BY WEAPON GRADE MICROWAVE FROM ANY DISTANCES, THE MICROWAVE - LASER PARTICLES ENTER INTO HEAD THROUGH EXTERNAL EAR CANALS, TRAVEL INTO MIDDLE EAR, TRANSIT TO INNER EAR THEN DIRECTLY ENTER INTO CNS AUDITARY CENTER INTERNAL ELECTRON NUCLEON CONSTRUCTIONS, POWERFUL LASER MICROWAVE FUNDAMENTAL PARTICLES DESTROY EVERY STRUCTURE IN THIS PATHWAYS, DEPENDING WHAT KIND AND WHAT DOSES MICROWAVE

ARE SHOT INTO HEAD, AND HOW MUCH AND HOW LONG CONTINUOUSLY MICROWAVE LASER CURRENTS WERE RUNNING INTO BRAIN.

AT EXTERNAL EAR CANAL THE COPIOUS EXCRETION OF THE CERUMEN IS THE MOST IMPORTANT DEFENSE SYSTEMS TO DIMINISH THE BIOHOSTILE SONIC PARTICLES DESTRUCTIONS, CERUMEN'S MAIN JOB IS NOT THE INSECT REPELLENT. CERUMEN MAIN FUNCTION IS TO DIMINSH HARSH POWERFUL DESTRUCTIVE VIBRATIONS AND ROARING POWERFUL CURRENTS OF MICROWAVE PARTICLE LASER CURRENTS, THROUGH DIMINISHING MICROWAVES POWERFUL VIBRATIONS AND FORCES THE CERUMEN PROTECTS THE MIDDLE EARS FINE BONY STRUCTURES FROM DISRUPTIONS AND PREVENTS TYMPANIC MEMBRANE RUPTURE, AND PREVENT THE CNS AND INNER EAR NEURAL CONSTRUCTIONS INJURIES AND DESTRUCTIONS. BECCAUSE COPIOUS CERUMEN EXCRETION BY EXTERNAL EAR COATS ALL AROUND THE EXTERNAL EAR CANAL WITH INSTANT EXCRETIONS AND COVER T.M. AS WELL, CERUMEN HELP TO REDUCE DESTRUCTIVE POWERS OF FORCEFUL VIBRATIONS AND MICROWAVE LASER SHEARS.

CERUMEN DIMINISH ROARING HARSH DESTRUCTIVE SHEARING FREQUENCIES OF LASER MICROWAVE PARTICLES, PREVENTS FROM TEARS AND DESTRUCTIONS OF FINE BIOLOGICAL TISSUES IN NEURAL PATH.

THE WAR SYNDROMES

SYMPTOMS AND SIGNS OF MICROWAVE WEAPON'S BRAIN INJURIES

SYMPTOMS AND SIGNS OF CNS ELECTRON - NUCLEON INJURIES ARE SIMILAR TO CNS CELLS INJURIES

SHOOTING WITH POTENT HIGH POWER INDUSTRIAL LASER-MICROWAVE WEAPONS ON HEAD OF INDIVIDUALS WITH BORDER LINE HEALTH CONDITIONS PRODUCE CONTINUOUS STIFF NECK, NAUCEA, VOMITING HEADACHES SIMILAR TO MENINGEAL IRRITATION SIGNS AND SYMPTOMS,

BUT THE PARA-CLINICAL TEST RESULTS AT INTERNAL ATOM INJURIES ARE NON- CONCLUSIVE AND NORMAL LOOKING. MICROWAVE LASER SHOTS PRODUCE INTERNAL CNS ATOM MOLECULAR CONSTRUCTION CHANGES, ALSO THE CONSTRUCTION CHANGES OF CNS ELECTRONS NUCLEONS CAUSE BEHAVIOR CHANGES AS RESULT. IN ELDERLY AND BORDER LINE HEALTH INDIVIDUALS THE MICROWAVE CAN PRODUCE ATOM MUTATIONS (CHEMICAL FORMULARY CHANGES) AND PHYSICAL CHEMICAL BIOLOGICAL FUNCTION CHANGES. THE INDIVIDUALS WHO ARE YOUNG, STRONG, HEALTHY WITH GOOD BODY IMMUNITY, RESIST BEING KILLED BY MICROWAVE INSULTS.

BUT MOST INDIVIDUALS WITHOUT EXCEPTIONS SUFFER FROM VARIABLE DIFFERENT MICROWAVE CHRONIC COMPLICATIONS, AND PERMANENT DISABILITIES SUCH AS NERVE DEAFNESS, PERMANENT NON - STOP TINITUS, AND CHRONIC DIFFERENT DEGREES BRAIN DAMAGES.

THE SYMPTOMS AND COMPLAINTS THAT CAUSED BY MICROWAVE INJURIES, ARE SIMILAR TO THE SYMPTOMS AND COMPLAINTS WHICH PRODUCED THROUGH THE OTHER DISEASES OF BRAIN DAMGE, LIKE: MENINGEAL INFECTIONS TYPE COMPLAINTS, GUN SHOT PRODUCED SYMPTOMS, BLUNT INJURIES TYPE COMPLAINT, ETC.).

GENERALLY, MICROWAVE DOE'S NOT LEAVE DEMONSTRABLE SIGNS TO PROVE THE EVIDENCE OF MICROWAVE INJURIES IN BEHIND. THE MICROWAVE LASER INJURIES NOT PRODUCE CUTS, NO REDNESS ETC. THE INJURED BRAIN BY MICROWAVE NEED PROLOGED PROPHYLAXIS ANTIBIOTIC TREATMENTS. THE SIGNS AND SYMPTOMS OF THE RADIATION (PARTICLE) INJURED TISSUES ARE SIMILAR TO RADIATION WOUND, DO NOT HEAL NORMALLY AND SECONDARY INFECTIONS OCCUR WITH SERIOUS SIDE EFFECTS.

MICROWAVE INJURIES HEALS VERY SLOWLY OVER YEARS TO RECONSTRUCT ELECTRONS NUCLEONS CONSTRUCTIONS, ALSO THE HEALING PROCESS MAY NEVER OCCUR AS WELL AFTER LONG TIMES IN SOME CASES.

WHEN PATIENTS SHOT BY MICROWAVE, ALWAYS PATIENTS FEEL TRANSITION PATHS OF HARSH POWERFUL ROAMING MICROWAVE CURRENTS, PASSING THROUGH EXTERNAL EAR CANAL TOWARD THE AUDITARY CNS CENTERS, AND THE PATIENTS CAN SHOW WITH FINGER POINT THE PLACE OF PAIN OVER SKULL, WHICH IT CORRELATES TO CNS HEARING CENTER LOCATION AT BRAIN.

THE MICROWAVE DEAFNESS AND HARSH CONTINUOUS TINITUS ROARS WITH NO STOP ALWAYS IS THE FIRST

COMPLAINT THE PATIENT STATES WHEN SHOT BY MICROWAVE ON HEAD. PATIENTS MOSTLY IN SERIOUS INJURIES MANIFEST BRAIN INJURY SIGNS AND SYMPTOMS.

THE SYMPTONS IN SERIOUS CASES ARE HEADACHE, STIFF NECK, NAUSIA, VOMITING, CONFUSION, FEVER, REMARKABLE NECK RIGIDITY, GRADUAL HEARING LOSS, CONTINUOUS HARSH ROARING WITH NO STOP TWENTY - FOUR HOURS, ETC. BUT CHANGES IN XRAYS, LAB. TESTS NON SPECIFIC, MAY NOT HELP. THE MRI MAY NOT HELP TO DIAGNOSIS.

FACT ABOUT INTERNATIONAL MICROWAVE PARTICLE WARS & SYNDROMES

IN MODERN INTERNATIONAL WARS, SHOOTING WITH COMPUTER OPERATED INDUSTRIAL POWERFUL POTENT BIOCIDAL LIGHT LASER -MICROWAVE FUNDAMENTAL PARTICLES BULLETS AND SHOTS DIRECT ON HEAD AND OTHER VITAL OR NON VITAL ORGANS. OR UNSING ANY OTHER BIOHOSTILE PARTICLES, THESE WEAPONS OF MASS DESTRUCTIONS ALSO CONTAMINATE AIR, SPACE AND ATMOSPHERE WITH FLOATING OBNOXIOUS MICROWAVE LASER PARTICLE CLOUDS. AND THESE ARE WEAPONS OF MASS DESTRUCTIONS AND CRIMES AGAINST HUMANITY THAT DAMAGES ORDINARY PEOPLES, CHILDREN, WOMEN, MEN ALIKE AS THE FIGHTING SOLDIERS IN WAR FIELDS ALL. ORDINARY PEOPLE LIVING IN THOSE REGIONS ARE NOT AWARE ABOUT THE PARTICLES IN AIR, AND THEY ARE LIVING, BREATHING UNDER CONTAMINATED MICROWAVE

LASER GENOCIDAL RADIATIONS CONTAMINATED-ATMOSPHERE, WHICH HAS BEEN PRODUCED BY RULERS TO REDUCE THEIR POPULATIONS, ALSO RULING OVER DISABLED AND BRAIN DAMAGE POPULATIONS ARE EASIER THAN NORMAL SMART INDIVIDUALS. THEY CAN RULE DECADES AFTER DECADES OVER HEAD INJURED (BY: MW) INDIVIDUALS. IN MANY NATIONS OF THE WORLD NOW RULERS ARE RULING IN MANY YEARS OVER THAN ONE HALF CENTURY.

WHEN AIRBORN MICROWAVE- LASER PARTICLES INTER INTO AUDITARY CNS CENTERS INTERNAL ELECTRON-NUCLEON, DESTRUCT AND TEAR DOWN NORMAL E. T. S. Y. – F.P. – I.I. - PARTICLE CLOUD –COMPOUND CONSTRUCTIONS OF THE ATOMS OF CNS CELLS. THE SYMPTOMS AND SIGNS ARE SIMILAR TO THE THOSE SYMPTOMS AND SIGNS OF NEUROLOGICAL BRAIN INJURY DISEASES.

THE SOLDIERS AND NATIVE CITIZENS MAY SUFFER FROM ACUTE PARTICLE INJURIES SYNDROMES WHEN THEY ARE IN NEIGHBORHOOD OR IN FIELD, OR THEY MAY DEVELOPE CHRONIC COMPLICATIONS IN LATER TIMES.

Microwave Laser injuries of Abdomen & Thorax

Laser Microwave Particle Currents in Tissue

THE LASER MICROWAVE PARTICLE SHOTS INTO CHEST CAVITY AND ABDOMEN

PATTERNS OF MICROWAVE LASER FUNDAMENTAL PARTICLES TRANSIT INSIDE BODY TISSUES

INSIDE TISSUES THE LASER MICROWAVE FUNDAMENTAL PARTICLES CURRENTS FLOW IN STRAIGHT LINES, IN SEPARATE PARALLEL PARTICLE FLOWS AND APART FROM EACH OTHER.

INSIDE CHEST CAVITY BIOHOSTILE POWERFUL MICROWAVE PARTICLES DO NOT FLOW AS ONE SINGLE PARTICLE CURRENT FROM ONE SIDE TO OPPOSITE POINT, THE POWERFUL OBNOXIOUS MICROWAVE PARTICLES FLOW INSIDE THE CHEST AND ABDOMINAL CAVITY AS SEPARATE PARTICLE CURRENTS FROM EACH OTHER, IN MULTIPLE STRAIGHT LINE SEPARATE INDEPENDENT FUNDAMENTAL PARTICLE FLOWS.

THE BIOHOSTILE LASER MICROWAVE INTERNAL BODY FUNDAMENTAL PARTICLE CURRENTS OR MICROWAVE LASER PARTICLE FLOWS ARE SENSIBLE WHEN CROSSING INSIDE THE HEART, LUNG, LIVER, AND OTHER ORGANS.

THIS SENSIBILITIES OF INDUSTRIAL POWER BIOHOSTILE MICROWAVE PARTICLES, THROUGH SENSING THE

PARTICLES REVEALS, THAT THE BIOHOSTILE MICROWAVE PARTICLES CURRENTS ARE NOT SINGLE ONE STREAM PARTICLE FLOWS INSIDE BODY TISSUES, BUT BIOHOSTILE MICROWAVE PARTICLES CURRENTS FLOW THROUGH: MULTIPLE DIFFERENT PARALLEL CURRENTS ONE NEXT TO OTHER IN STRAIGHT LINES BETWEEN INTERANCE POINT TO CHEST, BELLY AND BODY, AND STAY SENSIBLE THROUGH OUT ALL INTERRNAL BODY TRANSIT TIME, ALL INTERNAL BODY TRAVEL PATH, UNTIL IT EXIT FROM OTHERSIDE OF BODY THROUGH EXIT POINT, ALL ARE SENSIBLE.

THE HUMAN SPECIES WILL SENSE PAINLESSLY THE BIOHOSTILE PARTICLES, WHEN THESE MICROWAVE PARTICLE CURRENTS SHOTS INTO HUMAN'S CHEST AND ABDOMEN AND PARTICLE CURRENTS TRAVELING INSIDE DIFFERENT THORACO-ABDOMINAL TISSUES ALL ARE CLEARLY SENSIBLE.

WHEN THE BIOHOSTILE MICROWAVE PARTICLE CURRENTS INTER INTO THE CHEST CAVITY, AND FLOWING AS SEPARATE MULTIPLE PARALLEL PARTICLE CURRENT FLOWS, IN STRAIGHT LINES FROM POINT OF INTERANCE TO THE CHEST WALL DIRECT TO THE EXIT POINT OF BIOHOSTILE PARTICLES EXIT POINT FROM OTHERSIDE TO OUTSIDE BODY.

THE HUMAN BODY WILL SENSE PAINLESSLY THE FLOW OF BIOHOSTILE PARTICLE CURRRENTS INSIDE DIFFERENT INTERNAL BODY ORGANS SUCH AS: HEART, LUNGS, LIVER, AND INSIDE ABDOMINAL ORGANS WHEN THE MICROWAVE CURRENTS FLOWING AND CROSSING DIFFERENT BODY TISSUES. FINALLY EXITING FROM THE OPPOSITE SIDES OF ENTERANCES TO OUTSIDE.

ABDOMINAL THORACIC INJURIES BY LASER MICROWAVE WEAPONS PARTICLE SHOTS

BIOHOSTILE MICROWAVE PARTICLES CURRENTS INTERACTIONS WITH BODY ORGANS – PCS

THE INDUSTRIAL BIOCIDAL LIGHT -LASER MICROWAVE FUNDAMENTAL PARTICLE CURRENTS INSIDE DIFFERENT BODY ORGANS TISSUES OF CHEST CAVITY, ABDOMINAL CAVITIES FLOW ALTOGETHER AS PAINLESS TRAVELING MULTIPLE SEPARATE PARALLEL FUNDAMENTAL PARTICLE CURRENT STREAMS WHICH EACH ONE STREAM PARTICLE CURRENT RUNS PARALLEL TO THE OTHER MICROWAVE LASER CURRENTS, ALL INSIDE ONE SINGLE MULTI CURRENT BUNDLE ALTOGETHER RUNS IN STRAIGHT LINES.

COMPUTER OPERATED, LIGHT LASER MICROWAVE BIOLETHAL INDUSTRIAL FUNDAMENTAL PARTICLES BULLET SHOTS ENTER INTO CHEST AND ABDOMINAL CAVITY FROM ONE SIDE OF ENTERANCE POINT, THESE BIOHOSTILE PARTICLES TRAVEL AT STRAIGHT LINE INSIDE OF A BUNDLES OF MULTI- PARALLEL CURRENTS ALTOGETHER, AND EXIT FROM OTHER OPPOSITE SIDE OF CHEST OR ABDOMIN FROM

EXIT POINTS, ALL PATHS ARE SENSIBLE PAINLESSLY FOR ANIMAL SENSES.

THE MICROWAVE PARTICLE FLOW INSIDE THE THORACIC CAVITY AND ABDOMINAL CAVITY TISSUES, ENTER INTO ONE BODY ORGAN TISSUES TRAVEL IN STRIGHT LINE IN BUNDLE OF DIFFERENT LASER CURRENTS, DO SOME INTERATIONS WITH MEDIA AS WELL WITH DIFFERENT TISSUES IN ITS CROSSING PATHS.

LASER MICROWAVE EXIST FROM ONE ORGAN'S TISSUES AND THEREAFTER ENTER INTO THE NEXT INTERNAL ABDOMINO THORACIC CAVITY OTHER ORGANS TISSUES IN STRAIGHT LINES FROM ONE POINT OF BODY ENTERANCE TO OPPOSITE SIDE OF EXIT POINT IN OTHER SIDE OF ABDOMEN OR CHEST CAVITY. THESE DIFFERENT PARALLEL STRAIGHT LINE PARTICLE FLOWS ALWAYS STAY AS A BUNDLE, AND DO NOT FLOW AS ONE LARGE SINGLE STREAM FLOW, OR DOES NOT FLOW AS SINGLE ONE LARGE CURRENT IN THORACIC CAVITY TISSUES. (DIFFERENT PARTICLE CURRENT FLOW PARALLEL TO OTHER, INSIDE MULTI FLOW BUNDLES) THIS PATTERN OF PARTICLE CURRENTS ALL SENSED CLEARLY BY VICTIMS.

AS BEFORE EXPLAINED, THE ALL ANIMALS, LIVING THINGS ORGAN'S FUNCTION OPERATES UNDER THEIR ORGAN-SPECIFIC PARTICLE INTELLIGENCE SYSTEM CENTERS, AND ALL ELECTRONS NUCLEONS POPULATIONS OF THE ORGAN IS CONNECTED TO EACH OTHER, THROUGH ORGAN SPECIFIC PARTICLE CIRCULATION SYSTEMS.

WHEN THE POWERFUL LETHAL DOSES BIOHOSTILE LIGHT LASER CODED MICROWAVE SHOTS INTERRING INTO

DIFFERENT THORACO-ABDOMINAL ORGANS AND SYSTEMS, DEFFINITELY THESE BIOHOSTILE PARTICLE CURRENTS WILL INTERFERE WITH THESE VITAL ORGANS PARTICLE CIRCULATION SYSTEMS (PCS), SPECIFICLY IN ELDERLY, AND BORDERLINE HEALTH CONDITION INDIVIDUALS.

IN ELDERLY AND BORDER LINE HEALTH CONDITIONS, THE BIOHOSTILE MICROWAVE PARTICLES PRODUCE SERIOUS IRREVERSIBLE MEDICAL COMPLICATIONS SUCH AS DYSRHYTHMIA, WHEN BIOHOSTILE LASER MICROWAVE PARTICLES INTER INTO HEART TISSUES AND CROSS CARDIAC ORGAN, CAN DISRRUPT CARDIAC ELECTRICAL PARTICLE CIRCULATION SYSTEMS OF THE HESS- INTELLIGENCE SYSTEM CENTER, THE HESS INTELLIGENCE SYSTEM CENTER THROUGH CARDIAC PARTICLE CIRCULATION SYSTEMS (P.C.S.) CONTROLS ALL INTER -CARDIAC ELECTRONS NUCLEONS TOTAL POPULATIONS NANO – FUNCTIONS. THE POWERFUL STRONG BIOHOSTILE MICROWAVE LASER PARTICLES DISRUPT CARDIAC PARTICLES CIRCULATIONS SYSTEMS FROM HESS CENTRAL INTELLIGENCE SYSTEM CENTERS, OR FROM ANY OTHER MAJOR BRANCHES TO ELECTRICAL FUNDAMENTAL PARTICLE CIRCULATION SYSTEMS OF THE HEART, ANY DISRUPTION OF HESS – P. C. S. CAN CAUSE CARDIAC FUNCTION DISORDERS AND COMPLICATIONS SUCH AS: DYSRHYTHMIA, C. H. F., M. I., CARDIAC ARREST, AND DEATHS.

THE PHENOMENONS SIMILAR TO HEART MAY OCCUR, IN OTHER INTRA-THORACIC ORGANS, IN LUNGS, AND OTHERS, UNDER LETHAL DOSES OF BIOHOSTILE MICROWAVE PARTICLES SHOTS. MICROWAVE EASILY CAN TEAR DOWN WHOLE OPERATTIONS OF DIFFERENT BODY ORGANS AND

SYSTEMS PARTICLE CIRCULATIONS SYSTEMS TOTAL OPERATIONS IN SECONDS, AND CAUSE ALL ORAGANS AND SYSTEMS TOTAL ELECTRONS NUCLEONS POPULATIONS NANO-FUNCTIONS DISAPPEAR, THAT MEANS NO BIOLOGICAL EXISTANCE AND FUNCTIONS AND DEATH.

The Weapons that their injuries Doesn't leave any traces from committed crimes

IN THE CASES OF BIOHOSTILE STRONG LASER - MICROWAVE SHOT - INJURIES ON HEAD WHEN SHOOTING DIRECT ON THE HEAD CAUSE IRREPAIRABLE DAMAGES FREQUENTLY PRODUCE HEADACHES, STIFF NECK, NAUSIA LOCAL TENDERNESS EXACTLY ON SKULL SIDE OVER THE INJURED CNS- ELECTRONS- NUCLEONS INJURED TISSUES AND LOW LEVEL FEVER ETC. ALL ARE DETECTABLE BY EXAMINERS,

IN CONTRAST THE CHEST INJURIES DOES NOT PRODUCE TOO MUCH SIGNS OR SYMPTOMS IN NORMAL YOUNG STRONG INDIVIDUALS, UNLESS THE OCCURENCES OF COMPLICATIONS SUCH AS ARRHYTHMIA, SUDDEN MI, CHF,

CARDIO-PULMONARY ARREST, DIFFERENT TYPES RHYTHM DISORDERS, ETC. OCCUR SPECIFICALLY IN BORDERLINE HEALTH CONDITION AND ELDERLY INDIVIDUALS, WHEN THEY GET SHOT AND WERE HIT BY POWERFUL INDUSTRIAL LIGHT LASER SUPRASONIC MICROWAVE STREAMS AND FUNDAMENTAL PARTICLE CURRENTS ON HEART AND VITAL CARDIOVASCULAR SYSTEM'S PARTICLE INTELLIGENCE SYSTEM CENTERS (PIS) AND CARDIOPULMONARY PARTICLE CIRCULATION SYSTEMS (PCS) CIRCUITS.

SHOOTING LASER MICROWAVE POWERFUL FUNDAMENTAL PARTICLE CURRENTS DIRECTLY TARGETING CARDIOPULMONARY, AND CARDIOVASCULAR PIS, PCS, IN ELDERLY CAN EASILY CAUSE DYSRHYTHMIA, CARDIOVASCULAR CONDUCTION AND PCS RHYTHM BLOCK, HEART ARREST. THE CARDIOVASCULAR AND CARDIOPULMONARY PCS AND PIS DISORDERS CAN CAUSE SECONDARY MI AND OTHER CARDIOPULMONARY COMPLICATIONS WHICH ARE NON DISTINGUISHABLE FROM NATURAL CARDIAC ARREST AND DEATH CAUSES. THE INDIVIDUAL KILLED BY LETHAL WEAPONS OF MASS DESTRUCTIONS, WITHOUT LEAVING ANY TRACE FROM THE CRIMES AND CRIMINALS.

IN THE HANDS OF WELL PAID AND TRAINED AGENTS, UNDER AUTOMATED COMPUTERIZED OPERATION OF WEAPONS OF MASS DESTRUCTION BY SHOOTING THE VICTIMS DIRECTLY BY LIGHT LASER MICROWAVE INDUSTRIAL OR MILITARY GRADE WEAPONS OF MASS DESTRUCTIONS CAN PRODUCE MASS MURDERS, MASS TORTURES AND INJURIES ATANY PLACE AND ANY TIME, EVEN INSIDE THE VICTIMS OWN HOMES, OWN BEDROOMS,

LIVING ROOMS DAY AND NIGHT UNDER COMPUTERIZED OPERATIONS.

BIOHOSTILE LASER MICROWAVE WEAPONS, AND CHEMICAL, NUCLEAR, BIOLOGICAL WEAPONS ALL ARE WEAPONS OF MASS DESTRUCTION. THIS BOOK IS AN INTRODUCTION FOR LASER MICROWAVE COMPUTERIZED OPERATION WEAPONS OF MASS DESTRUCTION, WHICH ARE WIDELY USED TO KILL ORDINARY CITIZENS AT THEIR HOMES AND DISABLE SOLDIERS AT WAR FIELDS AS WELL, WITHOUT KNOWLEDGE OF EITHER CITIZENS OR SOLDIERS.

THERAPY OF MICROWAVE INJURIES

PATHOLOGY OF CNS ELECTRONS, NUCLEON'S PARTICLE COMPOUND DAMAGES BY MICROWAVE

MAIN PATHOLOGY

THE MAIN PATHOLOGICAL DAMAGES OF WEAPONIZED MICROWAVE FUNDAMENTAL PARTICLE INJURIES OCCUR IN AUDITARY CNS ELECTON NUCLEON PARTICLE COMPOUND MOLECULAR CONSTRUCTIONS. MICROWAVE LASER WEAPONS POWERFUL BIOCIDAL FUNDAMENTAL PARTICLES MOSTLY DESTROY CNS S. Y. T. E.- FP – I.I. – P. Cl. - COMPOUND MOLECULAR CONSTRUCTIONS IN AUDITARY CNS ELECTRONS AND NUCLEONS.

THE POWERFUL BIOHOSTILE LIGHT LASER MICROWAVE FUNDAMENTAL PARTICLES CURRENTS FORCES DESTROY AND TEAR DOWN NORMAL S. Y. T. E. –F.P. – I.I. – P. Cl. – COMPOUNDS OF CNS AUDITARY CENTERS AS WELL AS THE ALL OTHER CNS CENTERS ELECTRONS NUCLEONS NORMAL PARTICLE COMPOUND COMPOUNDS. AND IN THEIR PLACE, NON LIVE BIOHOSTILE MICROWAVE PARTICLE COMPOUNDS CONSTRUCT HARMFUL PATOLOGICAL MOLECULAR CONSTRUCTIONS. WHICH HARSH CONTINUOUS ROARING MICROWAVE FUNDAMENTAL PARTICLES WITH NO INTRUPTIONS DISABLE THE INFLICTED PATIENTS LIFE MOSTLY.

THE NORMAL ELECTRON NUCLEON S. Y. E. T. –F.P. – I.I. – P. Cl. COMPOUND CONSTRUCTIONS OF AUDITARY CNS CENTER ATOMS, AS WELL AS THE OTHER CNS PARTICLE COMPOUND AND COMPOSITE STRUCTURES UNDER MICROWAVE LASER PARTICLE FORCE TEAR DOWN AND DISABLE THE BRAIN FUNCTIONS. THEREAFTER BIOHOSTILE MICROWAVE PARTICLES COMBINE WITH INTERNAL ELECTRON NUCLEON PARTICLE COMPOUNDS, UNDER A. S. I. FP. Mol. CIC. PRODUCE PERMENANTLY HARSH ROARING MICROWAVE PARTICLE COMPOUNDS INSIDE THE CNS ATOMS. WHICH PATIENTS ARE UNABLE TO HEAR AND UNDERSTAND, AS WELL AS NON- INTERRUPTED CONTINUOUS TINNITUS WITH NO STOP ARE THE RESULTS.

TREATMENT

THE THERAPY FOR THIS PROBLEM IS ELIMINATION AND REMOVING (RETRIEVAL) ROARING ABNORMAL DESTRUCTIVE NON LIVE MICROWAVE FUNDAMENTAL PARTICLES OUT OF CNS AUDITARY AS WELL AS OTHER CNS CENTERS ELECTRON NUCLEON MOLECULAR COMPOUNDS CONSTRUCTIONS. THROUGH DEGENERATIVE A. S. I. F.P. Mol. C.I.C.

FOLLOWING REMOVAL OF LASER MICROWAVE TO OUT OF CNS ELECTRONS NUCLEONS CONSTRUCTIONS, IN THEIR MICROWAVE MOLECULAR SITES NEED RECONSTRUCTION OF THE CNS ELECTRONS AND NUCLEONS CONSTRUCTIONS WITH NORMAL S. Y. E. T. –FP – I.I. – P. Cl. COMPOUND CONSTRUCTIONS. RE-IMPLANTING NORMAL S. Y. E. T. – FP – I.I. – P. Cl. - COMPOUNDS THROUGH REGENERATIVE A. S. I. F.P. – Mol. C.I.C. INSIDE THE AUDITARY CNS INTERNAL ELECTRONS- NUCLEON PARTICLE COMPOUNDS REGENESIS.

THIS IS RE- INSTALLATIONS AND RECONSTRUCTION OF CNS ELECTRONS NUCLEONS PARTICLE COMPOUNDS MOLECULAR CONSTRUCTIONS WITH NORMAL S.Y. T. E. – F.P. – I.I.- P. Cl. COMPOUNDS STRUCTURES. IN THE PLACE OF PREVIOUSLY RESIDING ROARING ABNORMAL MICROWAVE PARTICLE COMPOUND CONSTRUCTIONS.

TRYING TO STORE NORMAL PARTICLE CLOUDS, AND PRODUCE NORMAL PARTICLE CLOUD STORAGE, FOLOWING RETRIEVALS OF MICROWAVE PARTICLES TO OUT OF ATOM CONSTRUCTIONS.

THIS PROCEDURES REVERCE THE PREVIOUSLY PRODUCED NEURAL NONFUNCTIONAL DEAFNESS DISORDERS. AS WELL AS WITH REMOVAL OF HARSH MICROWAVE FROM CNS

ELECTRONS NUCLEONS COMPOUNDS, THE HARSH CONTINUOUS ROARING TINNITUS ALSO GRADUALLY WILL ELIMINATE AND DISAPPEAR.

ANOTHER GROUP ABNORMAL FUNDAMENTAL PARTICLE DISEASES

FUNDAMENTAL PARTICLE TRANSMITTED DISEASES

THE ANOTHER EXAMPLE FROM ABNORMAL FUNDAMENTAL PARTICLE CLOUD TRANSMITTED DISEASES ARE THE PSYCHIATRIC DISEASES, IN PSYCHIATRIC DISORDERS, THE ABNORMAL SONIC, LIGHT, ELECTRIC, THERMAL FUNDAMENTAL PARTICLE INFORMATION IMAGE PARTICLE CLOUD TRANSMIT FROM ABNORMAL PARTICLE CLOUD DONNER SICK INDIVIDUALS TO NORMAL PSYCHE CHILDRENS AND STUDENTS ARE THE CAUSE FOR PRODUCTIONS OF THE MENTAL DISORDERS (ABNORMAL PARTICLE CLOUD TRANSMITTED DISEASES).

ABNORMAL PARTICLE CLOUDS SUCH AS ABNORMALLY CONSTRUCTED S. Y. T. E. – FP – I.I. – P. Cl. TRANSMIT AIRBORNE FROM MENTALLY SICK INDIVIDUAL TO NORMAL RECIPIENTS INDIVIDUALS ARE THE CAUSES OF THE PSYCHIATRIC DISORDERS.

THE ABNORMAL S. E. Y. T. - FP – I. I. – P. Cl. UNDER ATTRACTIVE GRAVITON FORCES OF SENSORY ELECTRONS AND NUCLEONS ENTER INTO CNS ATOM'S SUBSYSTEM

UNITS CHEMICAL LAB.S OF NORMAL INDIVIDUALS, UNDER REGENERATIVE A. S. I. F. P. Mol. C.I.C. THE INCOMING ABNORNAL PARTICLE CLOUDS COMBINE WITH INTERNAL ELECTRON NUCLEON PARTICLE COMPOUNDS OF THE RECIPIENT CHILDREN, STUDENTS AND INDIVIDUALS. PRODUCE ABNORMAL S. Y. T. E. – FP – I.I. – Mol. COMP. CONSTRUCTIONS OF THE RECIPIENT KIDS AND INDIVIDUALS INTERNAL CNS ELECTRONS AND NUCLEON CONSTRUCTIONS.

THE INTERACTIONS OF THESE ABNORMAL S. Y. T. E. – FP – I.I. – P. Cl. WITH EACH OTHER UNDER A. S. I. FP. Mol. CIC. BETWEEN DIFFERENT CNS CENTER ATOMS AND CELLS PRODUCE ABNORMAL THOUGHT CURRENTS AND ABNORMAL PSYCHE IN RECIPIENTS. THE PRODUCED ABNORMAL THOUGHT CURRENTS AND PSYCHE OF DONNORS HAVE SIMILARITIES WITH THE RECIPIENTS. THE PSYCHOTICS TRANSMIT PSYCHOSIS TYPE PARTICLE CLOUDS, DEPRESSION EMIT AND TRANSMIT DEPPRESSIVE TYPE PARTICLE CLOUDS, NEUROTIC DO THE SAME. TRANSMISSION OF ABNORMAL S. Y. E. T. - FP– I.I. – P. Cl. CAUSE STRORAGE OF THESE ABNORMAL CLOUDS INSIDE THE RECIPIENT KIDS AND STUDENTS CNS ELECTRONS NUCLEONS.

TREATMENTS

FIRST PHASE TREATMENT OF CHOICE FOR ABNORMAL PARTICLE CLOUD CONSTRUCTIONS AND STORAGE SYSTEMS IS TO STOP SOURCES OF THE MORE CLOUD TRANSMISSIONS FROM ABNORMAL DONNORS TO NORMAL RECIPIENTS AT FIRST STEP.

THE SECOND PHASE ESSENTIAL TREATMENTS OF ABNORMAL PARTICLE TRANSMITTED DISEASES IS RETRIEVAL AND REMOVAL OF THE EXISTING ABNORMAL S. Y. T. E. – FP – I.I. - P Cl. COMPOUNDS CONSTRUCTIONS AND INTERNAL ATOM STORAGE SYSTEMS, AND TRANSMITTING ABNORMAL PARTICLE CLOUD COMPOUND CONSTRUCTION MOLECULES INTO OUT SIDE CNS ELECTRONS AND NUCLEONS CONSTRUCTIONS THROUGH THE DEGENERATIVE A. S. I. FP. Mol. CIC.

THE THIRD ESSENTIAL STEPS OF THE ABNORMAL PARTICLE TRANSMITTED DISEASES IS RECONSTRUCT AND REPROGRAM THE CNS ATOMS AND CELLS INTERNAL ELECTRON NUCLEONS PARTICLE COMPOUND CONSTRUCTIONS WITH NORMAL TRANSMISSION OF S. Y. T. E. –FP – I.I. – P. Cl. COMPOUND REGENSIS AND STRUCTURES AGAINS. WHICH INTERACTIONS OF THESE NEWLY NORMALLY CONSTRUCTED NORMAL PARTICLE COMPOUND CONSTRUCTIONS BETWEEN DIFFERENT CNS CENTER ATOMS AND CELLS PRODUCE NORMAL PSYCHE AND NORMAL THOUGHT CURRENTS. THE FOURTH PHASE IS MAINTAIN ALL ABOVE FOR THE REMAINING LIFE.

THE ABOVE PHENOMENON IS ELIMINATIONS AND RETRIEVAL OF ABONORMAL S. Y. E. T. – F.P. – I. I. – P. Cl. TO OUT OF CNS- ELECTRONS NUCLEONS PARTICLE CLOUDS COMPOUNDS CONSTRUCTIONS. AND IN THEIR PLACE REIMPLANT OR STORE AND RECONSTRUCT THE CNS INTERNAL ELECTRON NUCLEON PARTICLE COMPOUNDS WITH NORMAL S. Y. E. T. – F.P. – I. I. – P. Cl COMPOUNDS AGAIN, AND MAINTAIN ALL OF ABOVE FOR REST OF LIFE.

THIS PHENOMENON OF ELIMINATION OF ABNORMAL PARTICLE CLOUDS AND REMOVAL OF ABNORMAL S. Y. T. E. – F.P. – I.I. – P. Cl. TO OUT OF CNS – INTERNAL ELECTRON NUCLEON CONSTRUCTIONS AND REPLACING THEIR PLACE WITH NORMAL S. Y. T. E. – F.P. – I.I. – P. Cl. - COMPOUND CONSTRUCTIONS AGAIN, THIS IS TREATMENT OF CHOICE FOR PSYCHIATRIC OR PARTICLE CLOUD TRANSMITTED DISEASES.

EZZAT E. MAJD POUR, M.D.

BRIEF DISCOVERIES OF THE AUTHOR DURING HIS LIFE

Discoveries of Author during Pre-Med & College years:

---Author discovered Electrons, Neutrons, Proton and Atoms are constructed from Fundamental Particles Compounds, Particle Clouds, have Particle Intelligence System Centers, and Particle Circulation Systems. Author during these years discovered Fundamental Particles General Chemistry, Organic Chemistry, and Biochemistry Sciences. Author discovered Molecular Evolution, how constructed different Subjects of Universe, under A. S. I. FP. Mol. CIC.

DISCOVERIES OF THE AUTHOR DURING MEDICAL SCHOOL YEARS:

--------------------Author discovered exogenous Particle Cloud Circulation Systems and Indigenous Particle Cloud Circulation Systems, Abnormal constructed Particle Clouds, and Normally constructed Particle Clouds, Author discovered causes of Psyche Genesis, and thought Current Genesis, Psyche of Atoms, Electrons, Nucleons. Author discovered the causes of the psychiatric disorders are Particle Cloud Transmitted diseases.

AUTHOR DISCOVERED PLANTS CENTRAL INTELLIGENCE SYSTEMS, PLANTS SENSORY INTELLIGENCE SYSTEMS, AND PLANTS MOTOR INTELLIGENCE SYSTEM CENTERS, AND PLANTS WITHOUT ANY INTELLIGENCE SYSTEMS, ETC.

AUTHOR'S DISCOVERIES DURING NINE POSTGRADUATE YEARS, AND THEREAFTER:

AUTHOR DISCOVERED HUNDREDS OF OTHER NEW FINDINGS, WHICH NO ONE HAS DONE THESE DISCOVERIES BEFORE, THIS BOOK IS AN INTRODUCTION FOR THE AUTHORS NEW CREATIONS.

04 – 10 - 2120

EZZAT E. MAJD POUR, M.D.

ADDRESS: ezzatmajd@icloud.com

MMMMDESTRUCTIONS, I APPEAL BANNING OF THESE WEAPONS.

www.ingramcontent.com/pod-product-compliance
Lightning Source LLC
Chambersburg PA
CBHW050158230526
45470CB00001B/139